M. Faber H. Niemes G. Stephan

With the Cooperation of L. Freytag

Entropy, Environment and Resources

An Essay in Physico-Economics

Second Edition

Translated from the German by I. Pellengahr

With 33 Figures

Springer-Verlag

Berlin Heidelberg New York
London Paris Tokyo
Hong Kong Barcelona
Budapest

Authors
Prof. Dr. Malte Faber
Universität Heidelberg
Alfred-Weber-Institut für Sozial- und Staatswissenschaften
Grabengasse 14, D-69117 Heidelberg, FRG

Dr. Horst Niemes
GKW-Consult, Besselstr. 26, D-68219 Mannheim, FRG

Prof. Dr. Gunter Stephan
Universität Bern, Volkswirtschaftliches Institut
Abteilung für Angewandte Mikroökonomie
Gesellschaftsstr. 27, CH-3012 Bern, Switzerland

With the Cooperation of
Dipl.-Phys. Dr. Lutz Freytag
AT 71 MBE, Siemens
Gewerbepark D 80, D-93059 Regensburg, FRG

Translator
Dr. Ingo Pellengahr
Fahrgasse 29, D-60311 Frankfurt am Main, FRG

Title of the original German edition:
Lecture Notes in Economics and Mathematical Systems, Vol. 214
Entropie, Umweltschutz und Rohstoffverbrauch:
Eine naturwissenschaftlich ökonomische Untersuchung
© Springer-Verlag Berlin Heidelberg 1983
ISBN 3-540-12297-4/0-387-12297-4

ISBN 3-540-58984-8 Springer-Verlag Berlin Heidelberg
New York Tokyo
ISBN 3-540-18248-9 1st Edition Springer-Verlag Berlin Heidelberg
New York Tokyo

Data conversion, printing and binding: K. Triltsch, Würzburg
42/2202-543210 – Printed on acid-free paper

Preface to the Second Edition

This book has been used as a text in the Departments of Economics at the University of Heidelberg (FRG) during the last decade and the University of Bern (Switzerland) during the last seven years. We therefore were glad when Dr. Müller of Springer-Verlag offered to publish a soft cover version of the second edition, to make the next economically more accessible to students.

Heidelberg and Bern, January 1995

Malte Faber
Horst Niemes
Gunter Stephan

Preface to the First Edition

This book is one of the products of a three-year research project. Our objectives were:

- to apply neo-Austrian capital theory to long-run problems of environmental protection and resource use;
- to develop an approach that takes physical relationships into consideration;
- to narrow the gap between theory and practice in environmental economics.

For this purpose, we established three interrelated research programs. In the first of these we supplemented and generalized neo-Austrian capital theory (STEPHAN 1980, REISS 1981, FABER 1986). In the second which is presented in this volume we developed an interdisciplinary approach to natural resources. Using concepts and methods from thermodynamics we investigated environmental and resource problems and their interrelationships. Finally, in the third research program we authored the monograph „Umweltschutz und Input-Output-Analyse. Mit zwei Fallstudien aus der Wassergütewirtschaft" (Environmental Protection and Input-Output-Analysis. With Two Case Studies in Water-Quality Management) (FABER, NIEMES, and STEPHAN 1983). Here we first gave an extensive description of the development of the management, in particular of the water-quality management, of the waste treatment sector of the BASF AG., Ludwigshafen/FRG[1], in the seventies. In this we could resort to prior work done by NIEMES (1978, 1981). With the help of the information from these plants, as well as from the chemical industry, we constructed a dynamic input-output model of the environment. This we then applied to the water quality management of the state of Baden-Württemberg from 1971 to 1979.

---

[1] This is one of the three great chemical enterprises in the Federal Republic of Germany; 50,000 workers are employed.

Whereas the monograph mentioned is a purely economic analysis with both theoretical and empirical content, the present work is the attempt to develop an interdisciplinary approach. An extensive overview of the content will be given in the second part of our Introduction.

Thermodynamic methods are used to relate natural science to our economic considerations. An essential reason for this is that in this way the "directedness" of physical processes, which equally applies to every economic production process, could be explicitly taken into consideration. For instance, GEORGESCU-ROEGEN (1971:3) writes:

> "The significant fact for the economist is that the new science of thermodynamics began as a physics of economic value and, basically, can still be regarded as such. The Entropy Law itself emerges as the most economic in nature of all natural laws."

We use thermodynamics, in particular the notion of entropy, in order to establish relationships between the economic system and the environmental system.

It is well-known that interdisciplinary research is as urgently needed as it is difficult to put into practice (KOOPMANS 1979, FABER and PROOPS 1985). On the one hand, the time and the effort necessary to make ourselves familiar with physical relationships and to apply them to economic relationships often seemed insurmountable. On the other hand, more than once we were so violently criticized that we seriously considered abandoning our attempts. Finally, however, we were helped by the insight that an interdisciplinary procedure can entail mistakes, and that if one wants to undertake interdisciplinary work, one will need the courage to expose oneself to the critique not only of one's own, but also to that of other disciplines. At the last stage of our endeavor, Mr. Lutz FREYTAG, who holds a degree in physics from the University of Tübingen, helped us in re-arranging and re-writing the physical Part II of our analysis. For this reason we have mentioned his collaboration in the title.

We are aware of the fact that it is almost impossible to give an introduction to thermodynamics in a space as confined as that in this book. We have thus foregone completeness and confined ourselves to a relatively simple presentation of those physical ideas that we needed for the development of our theory. In this we have to accept that to the natural scientist some concepts or relationships may seem to be explained in an unduly simplified manner. We considered, however, this procedure to be advisable for our more social science oriented readership.

The first German edition of this book has been used three times as a graduate textbook in economics at the University of Heidelberg. However, this book is not easy to read, even though the mathematical requirements are not beyond the knowledge of a student of economics. Difficulties arise on the one hand because we analyze four- and five-sector models over many periods. In order to facilitate the understanding of production and consumption in multi-sector models for the student, we have chosen the following procedure in Heidelberg. We started the course with Chapter 4 and the first two Sections of Chapter 5 of the "Introduction to Modern Austrian Capital Theory" (FABER 1979). This, however, is only a helpful, but by no means a necessary condition for an understanding of the monograph at hand. For economic readers another difficulty stems from the fact that they are unacquainted with the physical relationships presented in Part II, Chapters 3 and 4. In order to reach these readers we chose a procedure which assumed practically no prior knowledge of physics. Moreover, one can understand the following Parts III and IV without having read Part II. The case will be reversed for readers with a stronger background in physical science. These will find Parts I, III, and IV not easy reading. We hope, however, that the reader senses that we took much pain in presenting our material. Thus we use many examples from reality, draw many graphs, develop all derivations step by step, and summarize more extensive formal relationships in appendices to the chapters concerned. Wherever it was possible, we confined ourselves to a graphical presentation.

This book is addressed primarily, but not exclusively, to social scientists; it is also addressed to physicists, engineers, students, and those dealing with interdisciplinary environmental problems. We thus hope that our book will be read also by natural scientists and technicians interested in economic relationships.

We thank many friends and collegues who have read parts of the manuscript and have given us much advice and valuable criticism. Out of these we would like to especially mention the natural scientists Mr. Dipl.-Physiker Dietrich BRUNE (Kernforschungszentrum Karlsruhe (Nuclear Research Center Karlsruhe)), Prof. Dr. Werner BÖGE (Heidelberg), Prof. Dr. Ulrich FÖRSTNER (Hamburg), Prof. Dr. Reiner KÜMMEL (Würzburg), MARTIN O'CONNOR, Msc, MA (Auckland, New Zealand), Prof. Dr. Udo SCHERZ (Berlin), Dr. Dieter SCHUMANN (Heidelberg), Prof. Dr. Upke-Walther SCHMINCKE (Aachen), and an unknown referee of the Springer Press. We also wish to mention the economists Prof. Dr. Friedrich BREYER (Hagen), Prof. Dr. Peter FLASCHEL (Bielefeld), Prof. Dr.

Klaus JAEGER (Berlin), Prof. Dr. Hans Jürgen JAKSCH (Heidelberg), Gernot KLEPPER, Ph.D. (Kiel), Prof. Dr. Wilhelm KRELLE (Bonn), Gabriel LOZADA, Ph.D. (Stanford), and especially Dr. Gerhard MAIER (Heidelberg) who has read and discussed with us all versions of the manuscript.

The first German edition appeared in 1983, as Lecture Notes in Economics and Mathematical Systems No. 214 of the Springer Press. It was revised and in parts enlarged for the English edition by Malte FABER and Gunter STEPHAN. Mr. Dipl.-Volkswirt Ingo PELLENGAHR has, although in contact with the autors, independently translated the text. John PROOPS, Ph.D. (Keele/U.K.) examined the entire manuscript and made numerous suggestions with regard to both style and content. Dipl.-Volkswirt Peter MICHAELIS proofread the finished text and compiled the subject index. Mrs. Brigitte RÜGER typed our manuscript.

A subject as controversial as ours requires special emphasis of the fact that all responsibility for the way in which we have taken advice and criticism into consideration remains with us.

We finally thank the Deutsche Forschungsgemeinschaft for the considerable financial support of our endeavor over a two-year period, without which it could not have been realized.

Heidelberg, September 1987

Malte FABER
Horst NIEMES
Gunter STEPHAN

Contents

0 Introduction: Starting Point, Objectives, and Content

0.1 Starting Point and Objectives

Environmental protection and resource use are closely interrelated, for generally it is the transformation of raw materials that causes pollutants. The problems in these two areas are of a long-run nature, whether they are considered separately or in their relationship to each other. We would like to explain this in more precise detail:

1. Pollutants can accumulate over time so that the stock of pollutants grows. This can happen so slowly that, for a long time, it remains unnoticed until damage occurs that is either irreparable, or necessitates very costly treatment. An example of this is the deposition of cadmium in the muds of the Neckar river near our University at Heidelberg. Formerly, these muds could be used as agricultural fertilizers and soil-conditioners. As the treatment of these muds is not yet possible for economic reasons, while their removal remains necessary to allow the river to be used for shipping purposes, large deposit sites are needed. These deposit sites will need to be maintained for a long time into the future. This type of waste material deposit can thus be seen as undesired capital goods, which consume services instead of producing them.
2. The ability of nature to treat certain pollutants in the course of time is similar to that of a waste treatment plant. Thus the environment can also be considered as possessing the characteristic of a capital good. These capital goods of nature can be kept intact and improved by investment. They can also be destroyed, however, by overintensive use. An example of this is a lake where the algae multiply at a rate that does not allow the lake to recuperate due to eutrophy.

3. If the resource deposits in the environmental sector are interpreted as stocks of capital goods, then their exploitation can be considered as the consumption of capital good services. The more resources are extracted, the greater may be the environmental damage. If environmental damage is to be reduced or even avoided, resources must be used to a considerable extent. For a given quantity of resource extraction per period, this entails that the quantity of consumption goods per period decreases. If consumption is to be maintained, however, greater quantities of resources have to be extracted.

These observations show the interdependence of the problems of resource use and environmental protection. They illustrate also the temporal structure of these problems. In the economics of the environment and of resources, as in capital theory, intertemporal relationships are at the fore. Thus we have chosen a capital-theoretic approach for the analysis of these problems; this is the neo-Austrian approach, which was developed by BERNHOLZ, FABER, JAKSCH, REISS, and STEPHAN in the seventies.[1] A first contribution to the investigation of questions of waste treatment with this capital-theoretic approach has been made by NIEMES (1981: Chap. 3).

It appears to us that Austrian capital theory is especially well suited for studying intertemporal dependencies, for it explicitly considers the temporal structure of the economy in general, and of production in particular. It is already a result of the limits on the existing stocks of resources that, without continued technical progress, the steady growth (steady state) of an economy cannot be maintained. In the long run, without continued technical progress not even a stationary state can be maintained (GEORGESCU-ROEGEN 1975:367–9). In the production sector, the substitution of energy sources entails the substitution of capital goods. The resulting innovation (in SCHUMPETER's sense, see GERYBADZE 1982) implies a replacement of techniques in the course of time and thus structural changes, whose temporal development in turn entails adjustment problems. The oil price hike of the seventies has clearly proven this.

---

[1] Overviews and comparisons with other capital theories can be found in FABER (1979, 1986), REISS (1981), and STEPHAN (1980, 1988).

Environment and resource problems are being investigated by scientists in different fields. The significance of these questions for society has led to more and more interdisciplinary groups being formed, for the representatives of one scientific discipline alone can investigate only partial aspects. This is also true for economists (cf. KOOPMANS 1979:1). For this reason we have attempted to develop an interdisciplinary procedure. Several circumstances suggested the use of thermodynamics and especially its Second Law (known as the Entropy Law). The seminal work in analyzing economic relationships in this fashion has been done by GEORGESCU-ROEGEN (1971).[2] In his work he shows again and again that time is an essential variable of economic activity. The irreversibility of the exploitation of resources is an important example. Irreversibility and thus time dependency is also an essential characteristic of thermodynamic processes. Every real economic activity is based on some physical process which lowers free energy. Man, however, can only dispose of free energy. The First Law says that energy can neither be created nor destroyed. Thc Second Law, however, says that in a closed thermodynamic system the free, disposable energy decreases while the non-disposable energy increases.

A measure of this non-disposable energy is the entropy of the system. It can also be said that free energy corresponds to low entropy. In every physical process, including biological processes, a portion of the free energy dissipates. Man can no longer dispose of it, i.e. it is bound up in the system. Low entropy is thus constantly transformed into high entropy. This process continues until all the free energy of a system is dissipated and thus transformed into bound energy. Once this state has been reached, the system is in a state of thermodynamic equilibrium. If this happens with heat energy, then this process leads to the same temperature prevailing throughout the system. There is no longer the possibility to transform heat into work.

The dissipation of free energy indicates that entropy can also be interpreted as a level of order. If the amount of free energy is high (a low level of entropy), then the level of order of the system

[2] It was only towards the end of our present endeavor that we realized the extent to which this is true and how close is the relationship to Austrian capital theory. Thus GEORGESCU-ROEGEN (1978:1027–9) has developed a recycling model which is very similar to the one presented by us in Chap. 7 below.

is also high. Examples of this are large stocks of coal, oil, and gas with high concentrations (low entropy). This relationship holds correspondingly for other resources.

In the present book we shall develop an integrated physico-economic approach with which both environmental protection and resource depletion can be investigated. For this purpose we shall establish a formal relationship between entropy and the resource concentration in deposit sites. The notion of entropy will only be used descriptively, however, when we investigate the environmental damage caused by pollutants. With this procedure we follow a task that was formulated by GEORGESCU-ROEGEN (1971:5):

> "And in fact, the merit of the introduction of entropy as a new variable of state lies precisely in the analytical simplification and unification achieved thereby."

In particular, we wish to attain the following objectives:

1. an integration of physical and economic aspects of resource use;
2. an investigation of interdependencies between environmental protection and resource use on the basis of our thermodynamic and capital-theoretic approach.

Finally, we would like to point out a special feature of our approach. In Chaps. 1, 2, 6, and 7 we derive optimality conditions by maximization of a welfare function. Our approach thus corresponds to that of a centrally planned economy. It is known, however, that this approach can be alternatively interpreted as a decentralized market economy (MALINVAUD 1953, KOOPMANS 1957). How a decentralization can be effected for our model was shown in detail by FABER (1979:85–6).

There are two reasons for the fact that we work with a centrally planned economy. For one thing, it is an essential simplification for the analysis if we have to consider the decision of a central planning agency only instead of the decisions of many agents in an economy.[3] For another, neoclassical theory – on which traditionally-oriented economics is based – is heavily price-

[3] The informational difficulties which exist in reality and the resulting disadvantages of a planned economy as compared with a market economy have been examined in detail by BERNHOLZ and BREYER (1984, esp. Chap. 5).

oriented. The value problem is thus in the foreground. This is plausible, since it is the essential task of the economist to correlate goods with "values". It is our opinion that the real side of the economy, on the other hand, has been too long neglected. What we mean is that the direct analysis of the technology, technological substitution and technological progress, and the development of the stocks of resources and capital goods should be paid more attention than has hitherto been the case.[4]

0.2 Content

A graphic overview of our procedure, in which all four parts of our analysis and their interrelationships are indicated, is given in Fig. 0.1.

In Part I of our analysis we confine ourselves to an economic viewpoint. In Chap. 1 we give an introduction to dynamic models, in which we predominantly treat problems of environmental protection and resource use. At the same time we thereby intend to introduce the reader to the complexity of the object of our investigation. For the environment is used as a supplier of resources, as a supplier of public goods, as a site for economic activity, and as a recipient of pollutants, all at the same time. We also have to take into consideration the interdependencies between these functions of the environment over the course of time.

Whereas in the first Chapter we exclusively describe neoclassical analyses, we turn to a neo-Austrian approach in the second Chapter. This approach allows us explicitly to model the temporal structure of inputs, outputs, and of the ecological processes. For this purpose we utilize the prior work done by Niemes (1981). Since it is our experience that the analytical investigation of the effects of transformations of production on the size of consumption and the quality of the environment are difficult to understand, we explain the relationships graphically in Figs. 2.2 to 2.5.

[4] In the language of activity analysis, the value problem is the dual problem of an economy while the real production of goods is the primal problem. Since every dual problem has a corresponding primal problem and since the solution of the one implies the solution of the other, it suffices in principle to analyze one of them. The solution of each of the two problems requires attention to be focussed on different circumstances.

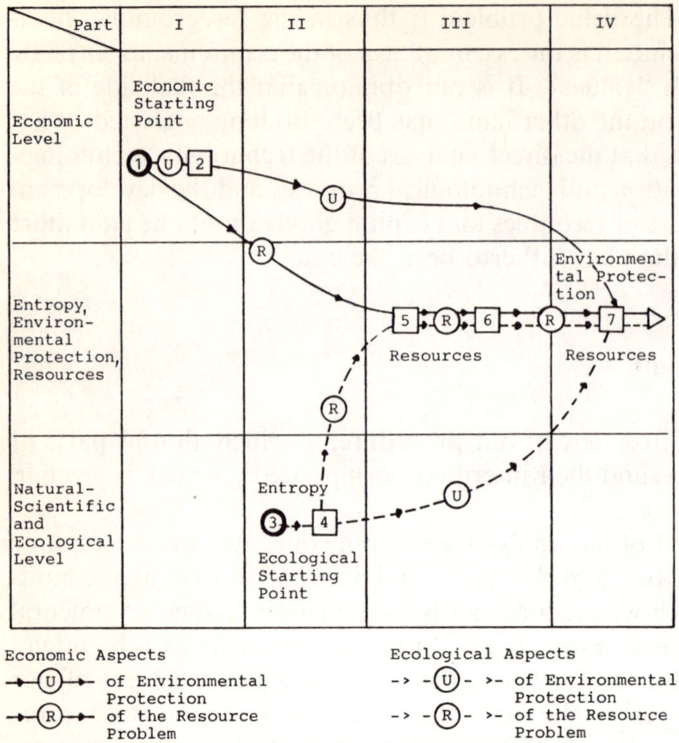

Fig. 0.1. Procedure and content

We then extend the model of NIEMES in several directions. In this way we obtain an encompassing environment-capital model, the optimality conditions of which we derive and analyze. In Part I, we restrict ourselves to economic aspects of environmental protection. We shall not return to the approach introduced in Chap. 2 until Part IV.

In Part II we develop our natural science starting point (cf. Fig. 0.1). Since the notion of entropy is very difficult to understand and at the same time of central importance for our approach, we devote the larger part of Chap. 3 to its introduction. It is well known that economics has been strongly influenced by classical mechanics for about a century. The development of thermodynamics since the beginning of the nineteenth century, however, has remained largely unnoticed by economists (cf.

MIROWSKI 1984). For this reason we have chosen to present in detail the thermodynamic relationships that are of importance for us. We hope that in this way we can highlight the difference between classical mechanics and thermodynamics. Thermodynamic processes are irreversible and thus process-dependent with respect to time; CLAUSIUS noticed this temporal aspect and introduced the notion of entropy, which stems from the Greek verb "turn over" (turn back, change). It can be argued that it was from classical mechanics that economists derived the attitude that economic processes are fully controllable once they have been fully described. Thus, in many models of growth theory the initial conditions and the growth rate suffice for a determination of the values of all variables at all times. The study of thermodynamic processes, however, shows that there are also uncontrollable variables in addition to controllable ones. Economists, of course, have noticed this, too. The following remark by LEONTIEF (1953:14), however, still applies to many economic analyses even today:

> "In principle at least, it has long been recognized that the ultimate determinants of the structural relationships which govern the operation of the economic system are to be sought outside the narrowly conceived domain of economic science. Notwithstanding their often expressed desire to cooperate with the adjoining disciplines economists have more often than not developed their own brand of psychology, their special versions of sociology, and their particular 'laws' of technology."

It remains to the critics to decide how far this is also true for our Part II. Here, we only wish to mention that Chap. 3 was written for economists and may – except for Sects. 3.5, 3.8 and 3.9 – be skipped by readers with a natural science background.

In Chap. 4 we use the notion of entropy to establish a relationship between economic activities and the environment. We shall interpret the separation process in the extraction of resources as a reversed diffusion process. Thereafter we shall derive relationships between resource quantities, resource concentration, entropy change, energy, and factor inputs. We shall use these in order to show how changes in the environment influence the economic production process. We shall establish the relationship between the economic system and the environment as a supplier of resources by way of the resource concentration. We thus directly utilize a variable of nature. With our entropy approach we extend the resource problem beyond the quantitative problem, by the inclusion of aspects of the distribution of resources within the environmental sector and the specific conditions within resource deposit sites.

These two aspects are explicitly taken into consideration in Part III, which deals with "The Use of Scarce Resources with Decreasing Resource Concentration". In Chap. 5 we integrate the resource problem into our capital-theoretic approach, using the same model structure as in Chap. 2. The common basic model is, however, extended by a resource sector. The waste treatment problem, on the other hand, remains temporarily outside of the analysis. We shall, however, be taking into consideration changes of resource quantities and concentration within the environmental sector. In Chap. 6 – similarly to Chap. 2 – we investigate the properties of our model by analyzing the effects of a rearrangement of production on the temporal distribution of the supply of consumption goods. In doing so, we are also interested in the replacement of techniques as a function of resource availability. We then derive optimality conditions for the temporal use of the environment as a supplier of resources. With the help of the variable 'resource concentration' we are able to show how the long-run increase of resource extraction costs can be explained as the result of technological and ecological conditions.

In Part IV we analyze interdependencies between environmental protection and resource use. For this purpose we join the environmental model of Chap. 2 with the resource model of Part III in a five sector model. With the examples of the recovery of resources from waste materials (recycling), and the controlled deposition of waste materials in the environmental sector, we show how our approach can be used to simultaneously investigate both environmental protection measures and resource use.

Part I

Environmental Protection

1 An Introduction to Dynamic Models of the Environment

As mentioned in the Introduction, we will start our investigations with a disaggregated environment-capital model. This was developed by NIEMES (1981: Chap. 3) from a neo-Austrian capital model. Even though some questions of environmental economics could be answered with this model, it did not quite suffice for an all-encompassing investigation of alternative waste treatment strategies. As we shall see in the course of our enquiry, a series of additional extensions are necessary.

Before turning to the extensions of this environment-capital model, however, we shall present some other dynamic approaches to the economic theory of resources and of the environment. In doing so, it is not our purpose to give an exhaustive overview of the literature. For this we refer the reader to MÄLER (1974), DASGUPTA and HEAL (1979), DASGUPTA (1982), FISHER (1981), VOIGT (1981), MÖLLER, OSTERKAMP and SCHNEIDER (1981, 1982), SIEBERT (1983) and MAIER (1984)[1]. Instead, we wish to explore them in order to demonstrate some open questions. At the same time we want to introduce our readers to the analysis of dynamic environmental and resource problems.

The models chosen for these purposes differ in the assumptions made and the questions posed. Thus some of them focus on problems of exhaustible natural resources, while others address questions of waste disposal or recycling.

Since all of these models work at a very high degree of abstraction, they do not yield any directly applicable results. However, they do allow insights into the intertemporal choice between the level of consumption and the quality of the environment. In many

[1] The relationship between the neoclassical theory of resources and the neo-Austrian approach is analyzed by MAIER (1984).

discussions the problem is often so narrowed as exclusively to address the choice between growth and environmental quality. This narrowing, however, is misplaced, for growth does not preclude environmental quality. Rather, we are asked not to invest exclusively in consumption goods industries any longer; a part of the new capital goods is needed for the treatment of emissions, the recycling of resources and the conservation – and possibly augmentation – of the regenerative and dissimilative[2] capacity of the environment. These considerations show clearly that environmental problems can be analyzed with the methods of capital theory.

1.1 The Relationships Between the Ecosystem and Its Economic Subsystem

Figure 1.1 makes it easier to compare the different models. In this Figure, those relationships between the ecosystem and its economic subsystem are represented that are essential for an investigation of the economics of resources and of the environment.[3]

The two arrows 0 and 0′ show that the ecosystem is an open system. Arrow 8 represents the process of individual or social appropriation of those environmental goods that are used for the construction of a production-consumption-cycle. Arrow 1 symbolizes that a part of the environmental goods is supplemented by human labor in the production sector of the economic subsystem to create a supply of goods corresponding to the demand by the economic agents (arrows 2′ and 3). Once the minimum of subsistence has been surpassed, a part of the produced goods can be channeled into the production sector P as capital goods (symbolized by the cycle made up of arrows 2 and 2′), while another part of those goods remains in the consumption sector (arrow 3). Not only consumption itself takes place in this sector; it also encompasses the decisions pertaining to consumption, investment, types of waste treatment, etc.[4] Both consumption and production cause

[2] Dissimilation is the degeneration of waste materials by nature.

[3] For a detailed description of this system schema see NIEMES (1981:16–21).

[4] Footnote see p. 13.

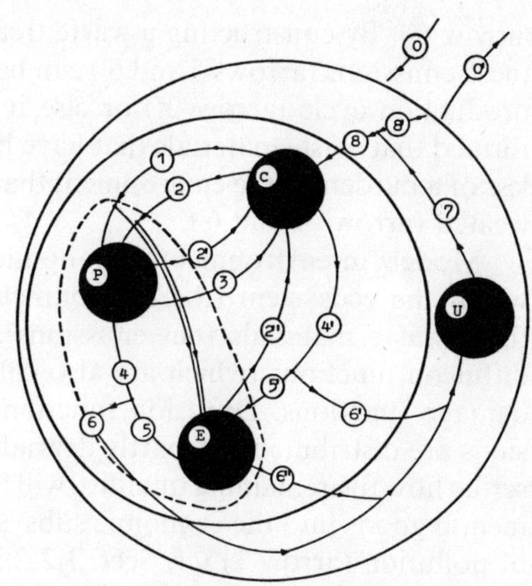

(1) Primary factors for production and waste treatment.
(2) Capital goods for production and waste treatment.
(2') Produced goods (capital goods) that remain in the consumption sector or that are utilized for production.
(2'') Goods that are recovered by recycling and rechannelled into the reproduction sector.
(3) Consumption goods.
(4) Waste materials of production (gross emissions).
(4') Waste materials of consumption.
(5) Waste materials of production to be treated.
(5') Waste materials of consumption to be treated.
(6) Non-treated waste materials of production.
(6') Non-treated waste materials of consumption.
(6'') Waste materials of treatment processes.
(7) Remaining waste materials which have not been eliminated by natural purification in the form of dissimilation or assimilation (immission).
(8') Resources in the environment.
(8) Resources affected by remaining waste materials (immissions).
(0) Flows from the total system into the ecosystem.
(0') Flows from the ecosystem into the total system.
C Consumption sector.
P Production sector.
E Waste treatment sector.
U Environmental sector.

Fig. 1.1. Relationships between the ecosystem and its economic subsystem

emissions, e.g. in the form of solid wastes or of waste heat (arrows 4 and 4′). If no special measures are taken, these emissions will be returned to the ecosystem (arrows 6 and 6′) or to the total system

[4] Instead of the representation chosen here, an additional capital sector might have been introduced. Our choice to represent it via the consumption sector C, however, is to emphasize that decisions concerning capital formation are not only determined by technical conditions and entrepreneurial choice but are subject to human choice in general.

(arrow 0′). By constructing a waste treatment sector E, a part of these emissions (arrows 5 and 5′) can be either channeled into the production cycle (arrow 2″) or else it can at least be so transformed that waste materials that have been treated (arrow 6″) are less of a burden on the environment than those that have not been treated (arrows 6 and 6′).

Models in environmental economics incorporate the way in which the ecosystem (the environmental sector U) deals with these waste materials (net emissions), for instance by way of diffusion functions (which are also called transfer functions) or damage functions. Diffusion functions describe how net emissions are distributed and partly degraded. Damage functions describe how the remaining quantity will impair the flow of environmental goods into the economic subsystem (arrow 8) in the form of pollution (arrow 7) (cf. Sect. 1.2.2.). Finally, arrow 8′ represents the original state of the flow of environmental goods into the economic subsystem; it includes the flow that is not affected by pollutants.

From this representation the different functions of the environment become apparent:

1. The environment is man's supplier of resources. We can distinguish between renewable and non-renewable[5] resources. The environment's ability to renew resources has the character of a capital good. By overuse or investment it can be negatively or positively affected by man.
2. The environment supplies man with many public (collective) goods such as forests, rivers or lakes, available for many different uses. As a consequence of the extraction of resources and the waste materials and harmful substances emitted, the quality of environmental goods is regularly diminished.[6]
3. The environment is the recipient of waste. Moreover, it is also capable of some waste treatment functions. This ability of the

[5] Since all resources have formed in the course of time, they are all renewable in principle. Strictly speaking the distinction is therefore between those resources that are renewable in the short-run and those that are only renewable in the long-run.
[6] Wastes are not always harmful. For examples, small amounts of food wastes in waters can have positive effects for the fauna and flora therein. Whether a waste material is regarded as harmful or useful depends in the last resort on the social preferences of the society. A distinction between wastes and harmful substances can therefore only be drawn after a particular problem has been specified.

environment to dissimilate waste has the property of a capital good. Man can influence this property negatively by the amount of waste and its constituents, and positively by investment in the environment.[7]

There are several interdependencies between these different types of uses. Thus if, e.g., pollutants are subjected to a recycling process in the waste treatment sector instead of being channeled into the environmental sector, valuable raw materials are recovered while the strain on the environment by pollution is simultaneously reduced. These mutual dependencies can, for instance, be described by the material balance approach (AYRES and KNEESE 1969, D'ARGE 1972, VICTOR 1972).

In order to explain this concept, let us look at the relationship between raw materials and pollutants in Fig. 1.1. In each period, the quantity of primary factors, ΔR (arrow 1 or arrow 8), equals the sum of the quantities of waste materials originating in the production sector (arrow 6), S_P, the consumption sector (arrow 6'), S_C, and the waste treatment sector (arrow 6''), S_E, plus the quantity of goods remaining within the economic subsystem, i.e. the produced capital goods I (arrow 2'):

$$\Delta R = (S_P + S_C + S_E) + I. \tag{1.1}$$

Comparability of the different substances listed in (1.1) is achieved by measuring each substance by its weight. Assuming that all waste materials are discharged into the environment, the quantity of substances reaching the economic subsystem from the environment within a time-period equals the net investment in that period

$$I = \Delta R - (S_P + S_C + S_E). \tag{1.1a}$$

This relationship shows that if raw materials are used intensively in the production of investment goods, e.g. by producing durable capital goods, the quantity of emissions will be smaller than in the opposite case. This has the consequence for environmental policy that if a particular level of investment I is aimed at in an economy

[7] SIEBERT (1978) lists its use as a location for economic activities as a fourth function of the environment.

and if emissions ($S_P + S_C + S_E$) are taxed, then the necessary extraction of resources ΔR will regularly be lower than if no taxes are levied; taxation results in an incentive to keep ($S_P + S_C + S_E$) as low as possible. For a given level of investment I there will thus be a tendency for the quantity of the primary factors which are needed for this investment and all economic activities to be lower. As a result of these reflections we recognize that charges on pollutants result not only in reduced damages and therefore in a better quality of the environment, but also in a reduction of the quantity of resources extracted for investment and consumption purposes. There is thus a close relationship between resources and pollutants. We shall investigate this relationship in the next Section.

1.2 Some Approaches to the Dynamic Analysis of the Use of the Environment

The examination of the three functions of the environment has shown that all three are mutually interdependent. We shall now explore some attempts that were made in the seventies to formalize and analyze the dynamic aspects of these functions. As mentioned above, in doing so we are not concerned with a detailed representation of the models, nor even with an exhaustive overview of the literature. We shall rather concentrate on those contributions that have either been paid particular attention in the literature, or that are of particular interest for our investigation.

In Sect. 1.2.1 we shall examine the environment as a source of raw materials and in Sect. 1.2.2 as a simultaneous recipient of pollutants and source of public goods. Since intertemporal models have a relatively complex structure, we shall at first ignore capital formation in order to render the modelling of the respective types of uses as simple as possible. Capital formation will be introduced in Sect. 1.2.3, in which all three functions of the environment will be analyzed. A model with natural purification will be followed by one with recycling.

1.2.1 The Environment as a Source of Resources

1.2.1.1 The Optimal Use of a Non-Renewable Resource

How can non-renewable resources be characterized? Simply in that they are limited in quantities and cannot be produced. This implies that extraction and consumption of a unit today involves some opportunity cost which depends upon the value of the resource at future dates, i.e. in determining the optimal use of a non-renewable resource over time, opportunity costs have to be taken into account. This observation directs attention to the questions: How do prices develop over time, if non-renewable resources are used optimally? What are the properties of the intertemporal opportunity costs of a non-renewable resource? In this Section we demonstrate how these questions can be answered in a simple continuous time framework. The starting point of our presentation is the approach of SOLOW (1974) and HEAL (1975). For further extensions see DASGUPTA and HEAL (1979).

Suppose there is a single producer-single consumer economy. We assume that both are price takers, i.e. they behave like agents in an economy with perfect competition. At each date of time t, amounts of a single non-renewable resource can be extracted which can be used for consumption without further inputs. The producer's total extraction costs $c(Q(t), R(t))$ are a function of both producer's output flow $Q(t)$ and the stock of resources $R(t)$ held at date t. Since the resource is non-producible, the stock $R(t)$ $(t > 0)$ is determined by the initial stock $R(0)$ and the total extraction prior to t at rate $Q(t)$:

$$R(t) = R(0) - \int_0^t Q(s)\, ds. \tag{1.2}$$

As is known from the literature (see for example HIRSHLEIFER 1972), a competitive equilibrium is achieved in this simplified economy if consumer's and producer's discounted net benefits are maximized over time. Net benefits at a single point of time are customarily measured as the difference between what consumers are willing to pay for a good and what it costs. In Fig. 1.2 this is the area between the consumer's willingness to pay function $p(Q(t))$ and the marginal extraction costs

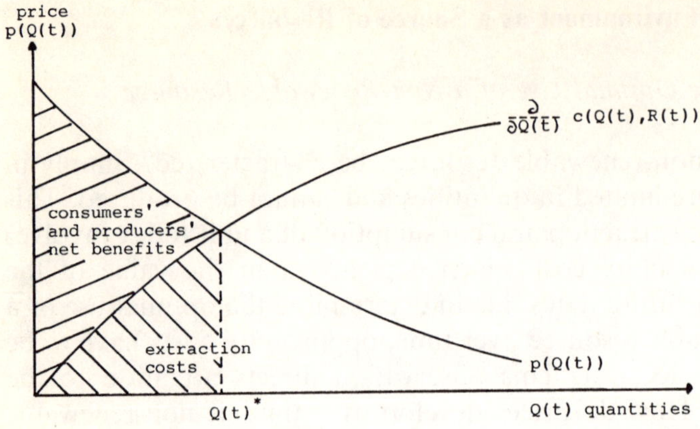

Fig. 1.2. Consumer's and producer's net benefits

$$\frac{\partial}{\partial Q(t)} c(Q(t), R(t)) \text{ between 0 and } Q^*(t).$$[8]

If $r\,(r > 0)$ denotes the interest rate, then the intertemporal problem can be expressed analytically as:

$$\max \int_0^{\infty} e^{-r \cdot t} \left[\int_0^{Q(t)} p(x)\,dx - c(Q(t), R(t)) \right] dt \tag{1.3}$$

subject to (by differentiation of (1.2))

$$\frac{dR(t)}{dt} = -Q(t). \tag{1.2a}$$

Applying the PONTRYAGIN maximum principle (see ARROW and KURZ 1970) we get the following Hamiltonian

$$H(Q(t), R(t), q(t), t) = e^{-rt} \left[\int_0^{Q(t)} p(x)\,dx - c(Q(t), R(t)) \right] - q(t)\,Q(t).$$

[8] Roughly, $p(Q(t))$ can be viewed as the first order derivative of the consumer's utility function. Formally, it is the inverse of the consumer's demand function.

By differentiating the Hamiltonian one obtains the first order optimality conditions

$$p(Q(t)) = \frac{\partial}{\partial Q(t)} c(Q(t), R(t)) + q(t) e^{rt} \quad \text{(each t)}, \quad (1.4)$$

$$\frac{dq(t)}{dt} = \frac{\partial}{\partial R(t)} c(Q(t), R(t)) \quad \text{(each t)}. \quad (1.5)$$

$p(Q(t))$ denotes the spot (market) or undiscounted price of the consumption good.

$q(t)$ is the so-called co-state variable. In contrast to $p(Q(t))$ the co-state variable $q(t)$ is a present value price.[9] It is attached to (1.2a), which describes the change in the stock of resources over time. In economic terms, $q(t)$ can be interpreted as the change in the optimal value of the objective function (1.3) corresponding to small changes in the constraint (1.2). In our case we are talking about the increase (decrease) in benefits at date t that would result from having one unit of resources more (less) available in the ground. But this is just the opportunity cost which accrues if one unit is extracted today. This opportunity cost is called royalty.

From this perspective (1.4) states that at an optimum[10] the price $p(Q(t))$ consumers are willing to pay has to be equal to the marginal costs of extraction $\frac{\partial}{\partial Q(t)} c(Q(t), R(t))$, plus the undiscounted royalty $q(t) e^{rt}$. (1.5) describes the development of the royalty over time, depending upon the size of the stock of resources.

To simplify the interpretation of (1.4) and (1.5), let us finally consider the pure royalty case which has figured prominently in the economic literature. Assume that there are no extraction costs at all, i.e. $c(Q(t), R(t)) = 0$. Hence (1.4) and (1.5) reduce to

$$p(Q(t)) = q(t) e^{rt} \quad (1.4a)$$

---

[9] The relationship between the present value price $q(t)$ and its corresponding spot price $\bar{q}(t)$ is as follows $\bar{q}(t) = q(t) e^{rt}$.

[10] Note that under the assumptions stated above each competitive equilibrium is Pareto-optimal.

and

$$\frac{dq(t)}{dt} = 0. \tag{1.5a}$$

Condition (1.5a) was first proved by HERFINDAHL (1955): The present value of the royalty does not change over time, or equivalently, the undiscounted royalty must rise at the rate of interest. Hence, we can set $q(t) = q_0$ and (1.4a) can be reformulated as $p(Q(t)) = q_0 e^{rt}$. By differentiating the last formula with respect to t we obtain

$$\frac{dp(t)/dt}{p} = r. \tag{1.4b}$$

Result (1.4b) is due to HOTELLING (1931) and states that under competitive conditions a non-renewable resource is used optimally over time, if its price rises at the rate of interest.

1.2.1.2 The Optimal Use of a Renewable Resource

PLOURDE (1970) developed a one-sector model with a renewable resource in which there is neither a production nor a waste treatment sector.

The economic subsystem in Fig. 1.1 thus consists of a consumption sector only. The resource may be used as a consumption good without the employment of labor. The starting point of the analysis is a flow equation, taken by PLOURDE (1970) from the literature on mathematical biology (LOTKA 1956). At every time t it determines the renewal rate of the resource dR(t)/dt as a function of its stock R(t):

$$dR(t)/dt = \lambda R(t) - \varepsilon R^2(t);$$

λ and ε are given parameters, whose sizes are determined by the type of resource. These parameters will differ substantially for different resources, e.g. for a stock of fish or a stock of forest. The function of the renewal rate is depicted in Fig. 1.3 for non-negative values.

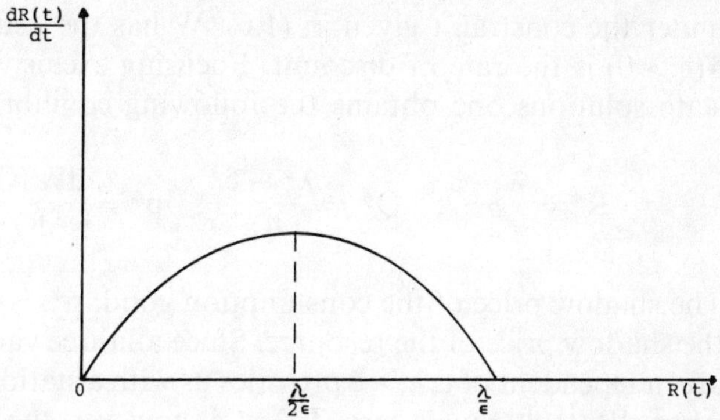

Fig. 1.3. Relationship between a stock of a renewable resource and its change

As long as the stock R (t) is smaller than λ/2 ε, the rate of growth increases, thereby augmenting the stock quickly. If R (t) is greater than λ/2 ε, the stock grows at a decreasing rate. From R (t) = λ/ε the rate of growth becomes negative. The settlement of population in new areas is an example of this relationship. While at first the population increases, as with the stock of fish in a newly constructed lake, it reaches a maximum due to the resulting shortage of nutrition. It will eventually decline if the nutritional resources are overused.[11]

If at any time t a part Q (t) of the stock is used for consumption, then the rate of change of the resource is

$$dR(t)/dt = \lambda R(t) - \varepsilon R^2(t) - Q(t). \tag{1.6}$$

The system is in a stationary state if Q (t) equals $\lambda R(t) - \varepsilon R^2(t)$, the original rate of change. The stock R (t) and the consumption flow Q (t) are constant.

In order to determine the optimal stock of resources, R*, and the optimal level of consumption, Q*, PLOURDE (1970) maximizes the welfare function

$$W = \int_0^\infty W[Q(t)]\, e^{-\delta t}\, dt \tag{1.7}$$

[11] Alternative formulations have been used by CLARK (1976) and FISHER and PETERSON (1976).

under the constraint given in (1.6). W has the usual properties; $\delta\,(\delta > 0)$ is the rate of discount. Focusing exclusively on steady state solutions one obtains the following equilibrium values:[12]

$$R^* = \frac{\lambda - \delta}{2\varepsilon}, \quad Q^* = \frac{\lambda^2 - \delta^2}{4\varepsilon}, \quad p^* = \frac{dW(Q)}{dQ}. \tag{1.8}$$

The shadow price of the consumption good, p^*, is of course also the shadow price of the resource. Since all three variables in (1.8) are independent of t, $\lambda > \delta$ provides us with a stationary solution, when $dR(t)/dt$ equals zero. If $\lambda \leqq \delta$, however, the resources will be used up completely. In this case there exists no stationary solution with positive consumption.

Plourde (1970) shows that the solution (1.8) differs from the one corresponding to the maximally sustainable consumption Q^0 and stock of resources R^0 in each period. This is

$$R^0 = \frac{\lambda}{2\varepsilon}, \qquad Q^0 = \frac{\lambda^2}{4\varepsilon}.$$

One recognizes immediately that the difference between the two solutions can be traced to the rate of discount δ; only for $\delta = 0$ are the two solutions identical. This result corresponds to the well-known result of the theory of growth, i.e. that with impatience to consume (i.e. $\delta > 0$) the golden rule path[13] needs not to be optimal. Only with neutral time preference ($\delta = 0$), i.e. if there is neither impatience nor patience to consume, are the solutions identical (cf. e.g. Shell 1967). This assumption is unrealistic, however, because as a rule there is impatience to consume in reality ($\delta > 0$). – We note that some economists since Ramsay (1928) have argued that $\delta > 0$ is morally indefensible. We shall come back to this problem in Sect. 7.3 below.

Finally we want to note that Plourde's (1970) model really only analyzes the relationships between the regenerative power of the system (expressed in the form of Eq. (1.6) and the parameters λ and ε contained therein), the stock of resources, consumption,

[12] Plourde (1970) derives these values by using the maximum principle of Pontryagin.

[13] The golden rule path is the path that allows maximal per capita consumption.

and time preference (expressed by δ). How simple this structure is and what essential ingredients of reality are missing can be seen from Fig. 1.1. Ecological processes in the environmental sector produce a stock of resources (e.g. a stock of forest), a portion of which flows as a stream of environmental goods into the economic subsystem (arrow 8), which in turn only consists of the consumption sector C. At each moment t the only economic decision is the determination of the level of consumption Q(t) according to the welfare function. This level of consumption in turn impairs the ecological regenerative power of the environment and thereby its function as a supplier of resources.

1.2.2 The Environment as a Recipient of Pollutants and as a Supplier of Public Goods

In the previous Section we used the models of SOLOW (1974), HEAL (1975) and PLOURDE (1970) to demonstrate the function of the environment as a supplier of resources. We now turn to its two other uses: as a recipient of pollutants and as a supplier of public goods. These two functions of the environment are closely interrelated. We first consider under what conditions, if any, they can be treated separately. This is the case, e.g., if it is possible to deposit pollutants in the environment without affecting its quality, for in this case the characteristics of the public goods, which are supplied by the environment, would not be influenced at all. In reality, however, the dissimilative processes (i.e. the natural purification) of the environment do not suffice to treat all pollutants received quickly enough. Thus pollutants accumulate in the environment until they finally reach a level at which they influence the quality of the environment and thus also the public environmental goods. As a consequence both uses of the environment, as a recipient of pollutants and as a supplier of public consumption goods, are in reality closely interrelated. From the point of view of environmental economics, a separate treatment makes, as a rule, little sense.[14]

[14] This conclusion holds in reality. From a theoretical viewpoint a separate treatment of these two functions may, however, be useful if the concept of a public good is more narrowly defined.

In Sect. 1.2.2.1 we shall first use the approach of D'ARGE (1972) to illustrate how emissions stemming from consumption can lead to an overstrain of the natural waste treatment capacity of the environment. This is especially the case if both population and per capita consumption are growing. Using a second model by PLOURDE (1972), we show in the succeeding Sect. 1.2.2.2 that this overstrain on the natural purification capacity can be counteracted by artificial waste treatment, thereby either maintaining or even enhancing the quality of public environmental goods.

1.2.2.1 The Accumulation of Pollutants

D'ARGE's contribution has recieved particular attention in environmental economics because he explains the intertemporal relationships between the economic subsystem and and the environmental system (cf. Fig. 1.1) using so-called "material balances". As explained above in the context of the derivation of Eqs. (1.1) and (1.1a), it is here assumed that while the materials used in economic and ecological processes may be transformed, their quantities are unaffected. Even if the materials used in production were transformed into consumption goods without leaving any wastes, the materials contained in these goods would be discharged into the environment when these are consumed.

The material-balance approach of D'ARGE is surely a restrictive interpretation of the energy-conservation approach.[15] Yet it helps both in pointing out relationships between the economic subsystem and the environmental system and in transforming these into ecological constraints for optimization models of the environment. D'ARGE demonstrates this procedure when he constructs his diffusion and damage functions.

What is crucial for the determination of the strain on the environment is not so much the quantity of the pollutants emitted, but rather their distribution over, and thus their density in, a given area. The latter is captured in a diffusion function in which the wastes arising in production and consumption are

[15] As already mentioned above we shall further discuss the limitations of the material-balance-approach in Chap. 3.

related to the volume V of the respective area. D'ARGE assumes that in consuming one unit of the consumption good in period t the quantity

$$S^b(t) = g\,q(t)\,N(t)$$

is left as waste.[16] Since the material balance approach is here being used, consumption and wastes are registered in terms of quantities. q(t) represents per capita consumption and N(t) the population in period t. g is a constant and depends on the technology used; ecologically beneficial technologies yield a small g, while ecologically harmful technologies yield a large g.[17]

Let S(0) be the quantity of pollutants and let D(0) = S(0)/V be its density at the beginning of period 0. Under the assumption that there is no natural purification we obtain the following density of pollutants for the beginning of period t:

$$D(t) = \frac{g}{V} \sum_{j=0}^{t-1} q(j)\,N(j) + D(0).$$

If one goes on to assume, for the sake of simplicity, that the natural treatment capacity of the environment is either impeded or small, one easily sees that with a growing population the density of pollutants would increase very rapidly, especially if per capita consumption rises simultaneously. If population grows at rate α and per capita consumption at rate β, we obtain

$$N(t) = \alpha N(t-1),\ \alpha > 1, \text{ and } q(t) = \beta q(t-1),\ \beta > 1,$$

and thus

$$\begin{aligned} D(t) &= \frac{g}{V} \sum_{j=0}^{t-1} \alpha^j N(0)\,\beta^j q(0) + D(0) \qquad (1.9)\\ &= \frac{g}{V} N(0)\,q(0)\,\frac{(\alpha\beta)^t - 1}{\alpha\beta - 1} + D(0). \end{aligned}$$

---

[16] A first clue as to the order of magnitude of wastes in an industrial society is given by the quantity of sewage sludge in sewage installations per capita and year. In Switzerland this amounted to 370 kg in 1980 (FAHMI 1981).

[17] The waste per household in the Federal Republic of Germany per year was ca. 300 kg in 1986. g is mostly greater than 1 since, as a rule, the production of consumption goods already causes wastes and pollution.

Since $\alpha, \beta > 1$, D(t) rises exponentially. This hypothetical development shows clearly how important it is not only to conserve the dissimilative capacity of the environment, but also to enhance it by investment, if possible.[18] In view of the size of α and β, dissimilation will generally not suffice; simultaneous artificial treatment will be needed. Using Eq. (1.9) one can also recognize the cumulative character of the burden on the environment. Not only the concentration D(t), but also the stock of pollutants

$$S(t) = g N(0) q(0) \frac{(\alpha \beta)^t - 1}{\alpha \beta - 1} + S(0) = D(t) V$$

grows exponentially because V is constant.

1.2.2.2 An Optimization Model

While the article by PLOURDE (1970), dealt with in Sect. 1.2.1.2, inquired into the influence of the consumption sector on the regenerative capacity of the environment and thus on its function as a supplier of resources, the approach of D'ARGE (1972) shows how consumption impedes the environment via pollution. These approaches are therefore complementary. While the former considers the extraction of resources, the latter analyzes the accumulation of pollutants.

We shall now proceed in our investigation to abandon the simplifications assumed by D'ARGE (1972) and consider both natural and artificial waste treatment. Artificial treatment allows the control of waste accumulation and thus simultaneously influences the quality of public environmental goods. These relationships are considered in another model by PLOURDE (1972), in which there is a consumption good and a public environmental good.

Contrary to the first approach by PLOURDE (1970), dealt with in Sect. 1.2.1.2, this model contains one waste treatment sector E and one production sector P. The environment in its function as

[18] An example for such investment is the construction of artificial inclines into barrage weired water courses. These contribute to the enrichment of water with oxygen, which is needed in the solution of pollutants.

a supplier of resources, however, is not considered in this model. Contrary to D'ARGE, for simplicity PLOURDE (1972) utilizes not the concentration of pollutants D(t) but its stock. The pollutants resulting from the production of the quantity Q(t) of the consumption good (gross emissions) can be decreased or, if necessary, even completely eliminated by waste treatment activities. Any pollutants remaining (net emissions) are discharged into the environmental sector, where they are partially dissimilated by natural purification processes.

Since those pollutants that are not degraded by "artificial" or by natural purification impair the quality of the environment and thus the well-being of the population, they are considered as undesired goods in a social welfare function. Their quantity is denoted by S(t). In addition to the produced consumption good there is also in PLOURDE's (1972) Model a collective good (public good) that is provided by nature. In order to demonstrate the relationships between the pollutants S(t) and the public good, PLOURDE (1972) implicitly assumes that a certain quantity of the collective good is given by nature. The more of the undesired good, i.e. the pollutant, there is, the smaller is the quantity of the collective good provided by nature and the lower is therefore welfare, ceteris paribus[19]. The welfare function is assumed to be as follows:[20]

$$W = \int_0^\infty W[Q(t), S(t)]\, e^{-\delta t}\, dt. \tag{1.10}$$

The function W[Q(t), S(t)] has the usual properties. In this case this means that the positive marginal utility of the consumption good decreases with rising consumption, and the positive marginal disutility of the pollutant increases when its quantity rises.

PLOURDE assumes further that for both the production of the consumption good Q(t) and the treatment of wastes there is only one non-reproducible factor of production, namely labor. The

[19] In this approach damages are thus derived only implicitly. In the terminology of the economic theory of the environment this means that there is no explicit damage function to show the relationship between immissions and the quality of environmental protection. In Sect. 1.2.3.1, Eq. (1.18), however, a damage function will be derived explicitly.

[20] Note that consumption is a flow whereas pollution is a stock.

question thus is: How should the quantity of the unique input factor $L(t)$, which is exogenously given at every point in time, be distributed between the two sectors? Let $L_Q(t)$ be the quantity of labor that is used for the production of the consumption good, and $L_S(t)$ the quantity used for waste treatment. Thus

$$L(t) = L_Q(t) + L_S(t).$$

Since labor is the only factor of production, there is no capital formation and no use of capital goods in production and waste treatment. The production functions are thus, for the production sector

$$Q(t) = f_Q[L_Q(t)];$$

for the waste treatment sector

$$S^e(t) = f_S[L_S(t)].$$

The accumulation of pollutants may, however, occur in the environmental sector. A measure of this is the change in the stock of pollutants, dS/dt, as a consequence of production and both artificial and natural waste treatment. PLOURDE (1972) assumes for the quantities of pollutants caused by production and consumption that

$$S^b(t) = g\,Q(t).$$

In addition he assumes for simplicity that natural purification is proportional to the stock of pollutants $S(t)$, i.e. $\gamma S(t)$, $0 < \gamma \leqq 1$.[21] This yields the flow Eq.

$$dS/dt = g\,f_Q[L_Q(t)] - f_S[L_S(t)] - \gamma S(t).$$

[21] Were $\gamma = 0$ allowed, the possibility that there is no natural treatment would also be encompassed. Since there are no activities in the model which allow the removal of net emissions once they have been discharged into the environment, this borderline case implies an irreversible occurrence; for even if there is no further discharge of emissions, the environmental system cannot return to its prior state. Natural purification, is then a necessary condition for the decumulation of pollutants discharged into the environment. In reality, γ depends both on the pollution already existing in the environment and on the resources already extracted. The natural waste treatment capacity of water, e.g., depends both on how much waste it contains, and on the amount of water extracted for production or consumption purposes (cf. THOSS et al. 1975:280).

Maximization of the welfare function under the above stated constraints yields the following results[22] (PLOURDE 1972):

1. The ratio of the shadow price for the consumption good to that for the treatment of wastes derived by optimization must be equal to the ratio of the marginal utility of the consumption good to the marginal disutility of the public environmental good.
2. At every moment t, the factor of production L(t) must be distributed between consumption and waste treatment so that the values of the marginal products in both sectors are equal to the marginal cost of the factor in both sectors.
3. The optimal stock of pollutants S(t) is positive.
4. The higher the disutility of pollution, the more resources are channelled into the waste treatment sector.
5. The higher the rate of discount δ, i.e. the higher the preference for present over future goods, the larger the stock of pollutants S(t).
6. The higher is γ, the more pollutants will be purified naturally. Since natural purification was assumed proportional to the stock of pollutants S(t), the larger the stock, the more wastes will be treated in this fashion (i.e. without cost to society).

Let us finally note that the structure of PLOURDE's model can be represented by arrows 1, 3, 4, 5, 6, 7, and 8 in Fig. 1.1.

1.2.3 Environmental Models with Capital Accumulation

While the environmental models presented so far encompassed accumulation processes in the environmental sector, they neglected the temporal structure of the economic subsystem. None of the models contained, for example, capital formation and accumulation in the production sector. In the following we shall therefore present models in which not only the environmental sector, but also the production sector exhibit temporal change.

---

[22] These results are valid under the assumption of a welfare function which is additive for every point in time. PLOURDE (1972:120–1) utilizes the maximum principle for the derivation of the results. This maximizes the welfare function (1.10) under the given constraints. The LAGRANGEAN multipliers derived can be interpreted as prices.

This seems important to us for the following reasons. If a capital good is to be employed in the production of the consumption good, the stock of capital necessary for this purpose must first be built up over time. The resultant temporal structuring of the production sector is essentially arrived at by adding another flow equation for the stocks of capital goods in the production sector to that for the stocks of resources and pollutants in the environmental sector. In order to demonstrate the close relationship between the different forms of uses of the environmental sector, we shall also consider recycling. This will finally allow us to explore the relationships between all three functions of the environment.

In the following two subsections we shall present two models by MÄLER (1974:Chap.3), both of which contain capital accumulation. In the first model a part of the pollutants is treated naturally by dissimilation, while in the second model waste materials are partially utilized as a substance from which new raw materials are won. As mentioned above, recycling thus presents the possibility of improving the quality of the environment and at the same time easing the problem of resources.

1.2.3.1 Natural Purification

MÄLER (1974:62–73) assumes a homogeneous good which can be used either as a consumption good or as a capital good. This good is extracted from a non-renewable raw material whose initial stock is R (0). The stock present at time t is denoted by R (t). Labor L (t) and capital K (t) are also used in the production of the homogeneous good Y (t):

$$Y(t) = F[K(t), L(t)].$$

Let the production function F have the usual neoclassical properties with the first derivative for each of the two factors being positive, and the second derivatives negative. Assuming that the quantity of labor at any time t is constant and fully employed, over the finite planning horizon T, we obtain for the production function

$$Y(t) = f(K(t)), \tag{1.11}$$

where $f'(K(t)) > 0$ and $f''(K(t)) < 0$.

All of the raw material is transformed into the homogeneous good. Therefore the quantity Y (t) is equal to the quantity of the raw material which is employed in production at time t. The resource constraint for the whole planning horizon is thus

$$\int_0^T f[K(t)]\,dt \leqq R(0). \tag{1.12}$$

At every time t production Y (t) is allocated in part as a non-durable consumption good Q (t) and in part as gross investment $I^b(t)$

$$Y(t) = Q(t) + I^b(t). \tag{1.13}$$

Let this division be irreversible. If it is further assumed that the stock of capital deteriorates at a rate c, we obtain for the rate of change of the stock of capital at time t

$$\frac{dK(t)}{dt} = I^n(t) = I^b(t) - cK(t), \tag{1.14}$$

and, as a result of the irreversible allocation mentioned above, for consumption

$$Q(t) \leqq Y(t).$$

For the stock of capital this means that decumulation cannot be smaller than $-cK(t)$:

$$I^n(t) + cK(t) \geqq 0. \tag{1.15}$$

Finally, we note that this model is based on the material balance approach. Thus the quantity on the left hand-side of (1.12) is equal to the quantity of resources that is extracted and transformed into produced goods during the whole time-span of T periods. Therefore MÄLER (1974) assumes that goods turn into waste materials when they are consumed. For simplicity it is further assumed that this happens without a time lag. Since the same applies to the deterioration of the capital good, we obtain for the total flow of waste materials, i.e. the gross emissions, at time t

$$S^b(t) = Q(t) + cK(t). \tag{1.16}$$

If it is further assumed that there is no artificial waste treatment and that the quantity of pollutants treated by natural purification is proportional to the quantity of pollutants accumulated in the environmental sector to date t, S(t), the following flow equation describes the latter's rate of change:

$$dS(t)/dt = S^b(t) - \gamma S(t) = Q(t) + cK(t) - \gamma S(t), \tag{1.17}$$

where $0 < \gamma \leqq 1$.

Whereas PLOURDE (1972) uses the quantity of pollutants directly in the welfare function (1.10), MÄLER (1974:64–5) utilizes a damage function which specifies the relationship between the stock of pollutants S(t) present in the environmental sector and the quality of the environmental good.[23] For the formulation of the damage function he utilizes the measure of concentration D(t) explained above in (1.9); the more pollution there is per unit volume, the lower is the "degree of environmental purity" and therefore the poorer is the quality of the environment U(t). This relationship may be expressed by the damage function

$$U(t) = 1 - S(t)/V = 1 - D(t). \tag{1.18}$$

Since $S(t) \leqq V$, i.e. the total quantity of pollutants may be less than or at most equal to the full volume V, the value of U(t) lies between zero and one. If U(t) = 1, the quality of the environment is at a maximum.

Before turning to the optimization of the welfare function we shall summarize the structure of this model with the help of Fig. 1.1. The raw materials (arrow 8) are appropriated in the consumption sector C and are channelled into the production sector P (arrow 1). Here the raw material is treated according to the production function (1.11) and then rechannelled into the

[23] This second environmental good may be a private or a collective good.

consumption sector (arrows 2′ and 3). Part of the raw material is consumed while the rest is used to increase the stock of capital K in the production sector (arrow 2). This latter part thus corresponds to investment. The waste materials of production, in the form of the deterioration of the stock of capital (arrows 4 and 6) and of consumption (arrows 4′ and 6′), are directly channelled into the environment, where they are subjected to natural purification. The remainder of them changes the stock of pollutants and impairs the quality of the environmental good according to (1.18). The lower quality environmental good is then received in the consumption sector (arrows 7 and 8). All relationships represented in Fig. 1.1, except those to the waste treatment sector E, are thus contained in this model.

Whereas PLOURDE (1972) considers the pollutants negatively in the welfare function, the degree of purity of the environmental good here enters positively:

$$W = \int_0^T W[Q(t), U(t)]\, e^{-\delta t}\, dt. \qquad (1.19)$$

Maximization of this welfare function (1.19) under constraints (1.11) to (1.18) yields the optimal temporal development of consumption, of environmental quality and the accumulation of pollutants, of investment and the capital stock, and of resource use and the decumulation of the stock of resources. MÄLER (1974:71) shows that under certain assumptions the optimal path has a turnpike characteristic. This means that for the greater part of the planning horizon it is close to a steady state solution.[24] The assumptions are

1. that the time horizon is long enough and
2. that the stocks of capital and of resources, both at the beginning and at the end of the planning horizon, conform roughly with the corresponding steady state equilibrium.

Were these assumptions met then politicians could strive for the optimal path of development with decision rules that are relatively simple to handle. For if the values of that steady state that is close to the optimal development are known, then the

[24] All variables change at the same rate over all periods.

capital stock actually present must be increased if its initial stock is smaller than that of the corresponding steady state, and vice versa; if the environmental quality is worse than that of the steady state, then waste discharges must be reduced. We may further conclude that, disregarding the initial and final periods, the turnpike characteristic ensures that the optimal path corresponds closely to a steady state. Thus one can very simply examine changes in the parameters of the model by comparing the corresponding steady state solutions. Since these are fully characterized by the initial stocks of resources, capital, and pollutants, and their respective rates of change which are constant over time, a comparison may be confined to these quantities. An increase of the stock of resources, e.g. as a result of a new discovery, allows the production of a larger quantity of the consumption good. Yet this necessitates a larger capital stock. Rising consumption and additional deterioration of the capital good increase the quantity of pollutants discharged into the environment (see (1.17)); as a consequence, the quality of the environment deteriorates. This last result depends, of course, on the fact that there is no artificial waste treatment in the model. Conclusions are reversed if the society decides to lengthen the planning horizon T, for in this case the quantity of the resource that can be used on average per period decreases.

1.2.3.2 Recycling

In the preceding model the quality of the environmental good could only be influenced indirectly via the size of the flow of consumption goods as well as the flow of materials due to the deterioration of the capital good and the resultant flow of waste materials. We are now going to supplement this model by the inclusion of a waste treatment sector.

The discharge of unpurified waste materials into the environment not only has the disadvantage that the quality of the environmental good deteriorates, but also that a possible source of resources remains unused. The recovery of resources during waste treatment, i.e. recycling, however, meets both disadvantages simultaneously. From the waste materials to be treated, resources are recovered and thus the burden that is put on the environment is smaller. Therefore recycling is a particularly suitable form of waste treatment.

The model of the preceding Section is supplemented by the following assumptions (cf. MÄLER 1974:87–94). The division of production is no longer exclusively between consumption and capital investment, for now there is also investment in the treatment of wastes. Total investment is thus divided between the production sector P and a newly added waste treatment sector E, the recycling sector. We therefore obtain for gross investment

$$I^b(t) = I^b_P(t) + I^b_E(t),$$

and analogously for the stocks of capital goods

$$K(t) = K_P(t) + K_E(t). \tag{1.20}$$

These distributions are irreversible. For the division of production, (1.13) is replaced by

$$f[K_P(t)] = Q(t) + I^b_P(t) + I^b_E(t).$$

In order that the present model be as similar as possible to the preceding one, MÄLER (1974) assumes that in the waste treatment sector the raw materials can be recovered only with the capital good and that the rate of deterioration is the same as in the production sector, $c = c_P = c_E$. Therefore

$$I^b_P(t) + I^b_E(t) = I^n_P(t) + I^n_E(t) + c[K_P(t) + K_E(t)]. \tag{1.21}$$

We now turn to the production function for the waste treatment sector. In order to make it conform as closely as possible to reality, MÄLER leaves the narrow confines of his model. In reality the flow of wastes does not consist of only one type of material, but contains many components. For each of these components it is assumed that the larger the flow of wastes, the more easily can a given quantity of a raw material R_E be recovered. Denoting the former by S^b, we obtain for the state of technology presently known

$$S^b > R_E.$$

The inequality holds as it is impossible to transform the whole flow of waste materials into resources. Let the quantity of the capital good that is needed for the recovery of R_E units of the resource from a flow of waste materials S^b be

$$K_E = h(R_E, S^b), \tag{1.22}$$

where h is a homogeneous function of degree 1. This means that the same quantity of the capital good is needed per unit of the resource if the proportion of R_E to S^b remains constant. If the proportion increases, more of the capital good is needed per unit of the resource.

The amount of pollutants discharged from the recycling sector into the environment equals net emissions $S^b - R_E$. Due to (1.16) and (1.20) one thus obtains

$$S^n = S^b - R_E = Q + c(K_p + K_E) - R_E.$$

Taking into consideration net emissions and natural purification, instead of (1.17) we now obtain for the rate of change of the quantity of pollutants in the environment

$$dS(t)/dt = Q(t) + c[K_p(t) + K_E(t)] - R_E(t) - \gamma S(t). \tag{1.23}$$

The recovery of the resources loosens the restriction concerning the resources present in the environment; instead of (1.12) the constraint is now

$$\int_0^s \{f[K_p(t)] - R_E(t)\} \, dt \leqq R(0), \quad \text{for every } s \leqq T. \tag{1.24}$$

In order to determine the optimal solution of this model the welfare function (1.19) must again be maximized under the restrictions mentioned above, including (1.18).[25]

A comparison of the solutions to both models presented in Sect. 1.2.3 yields the following results. If one sacrifices a part of

[25] Besides these restrictions, initial and final stocks of the capital goods should also be taken into consideration (MÄLER 1974:89).

consumption in an initial phase and uses this part for the formation of capital in the waste treatment sector, one may, as a result of the recycling procedure and after a certain time, be able to consume more and enjoy a better quality of the environmental good than would be the case without recycling. Whether such a stock of capital will be accumulated in the recycling sector depends, above all, on the rate of time preference of the society. In any case, recycling of resources may loosen the resource constraint.

A last glance at Fig. 1.1 shows that this model contains all of the relationships depicted there. Besides the consumption sector, the production sector and the environmental sector, we now additionally consider a waste treatment sector in which raw materials for production (arrow 2″) are recovered from a flow of waste materials S^b. Arrows 4 and 4′ denote the respective quantities of waste materials $c(K_P + K_E)$ and Q. The resources that are recovered are channelled into the consumption sector (arrow 2″) and there enlarge the flow of environmental goods (arrow 1). The remaining waste materials are discharged from the waste treatment sector into the environment (arrow 6″).

2 A Disaggregated Environment-Capital Model

In the first Chapter we provided an introduction to dynamic relationships in resource and environmental economics. In Sects. 1.2.1 and 1.2.2 we demonstrated the temporal structure of the three uses of the environment for mankind, simultaneously showing how all three functions are interrelated. In Sect. 1.2.3 we additionally considered the temporal structure of production by modelling both the production sector and the environmental sector with capital formation.

Since 1970 a number of papers dealing with dynamic environmental problems have appeared[1]; however in these their respective authors have used a procedure in which the temporal structure of the relationships is unfortunately often not given the prominence it deserves. An example is the model by MÄLER (1974) which was presented at the end of the preceding Chapter (Sect. 1.2.3.2). Inputs into production were there transformed into outputs without a temporal lag. The same applied to consumption goods, which were immediately transformed into waste materials while being consumed and, simultaneously impaired the quality of the environmental good via the quantity of the stock of waste materials. NIEMES (1981), on the other hand, chose a procedure which explicitly considered the temporal structure both of inputs and outputs and of the ecological processes. He did so on the basis of a variant of neo-Austrian capital theory that was developed by BERNHOLZ, FABER, JAKSCH, REISS and STEPHAN

[1] Beside those already mentioned see KEELER, SPENCE and ZECKHAUSER (1971), FORSTER (1972), SMITH (1972), OCKER (1974), FORSUND (1975), FORSTER (1975), THOSS et al. (1975), BENDER (1976), GRUVER (1976), STEPHENS (1976), FORSTER (1977), COMOLLI (1977), MAGAT (1978), BUCHHOLZ and CANSIER (1980) and JAEGER (1980).

in the seventies.[2] In addition, this approach allows the analysis of structural changes caused by innovations of new capital goods. Such processes can at present be observed particularly frequently in the waste treatment and the recycling sectors since these are subject to rapid expansion. Thus the BASF in Ludwigshafen/FRG e.g. invested approximately 500 million DM between 1965 and 1974 in order to purify its effluents (FABER, NIEMES and STEPHAN 1983:30–32). If one further adds the lengthy planning of these measures which began in 1959, the whole time span between the beginning of the development (time of initial input) and the time of the first purification of effluents (time of first output) is fifteen years. This example shows both the importance of innovations in the waste treatment sector and that of the vertical time structure of production. At the same time it demonstrates that developments in reality are not characterized by proportional growth (steady state), as is often assumed, but by structural change.

In the following Sect. 2.1 we shall summarize the model and the results of NIEMES's (1981:43–101) detailed analysis verbally and graphically. In Sect. 2.1.1 we describe the technology of his model and in 2.1.2 the temporal structure of the diffusion process and the damage function, i.e. the transformation of emissions into immissions and its influences on the flow of environmental goods. In Sect. 2.1.3 the effects of small changes in the production of consumption goods will be described. We shall see that marginal changes which take place during one period have effects that last for many periods. In this way we hope to show more clearly than was possible in the preceding Chapter how the different sectors are temporally interrelated. In Sect. 2.2 we shall consider the limits of the model by NIEMES. We shall then extend our model in order to deal with questions that had not been taken into account in our approach. The following extensions will be made. Firstly, the formation of capital in the waste treatment sector (Sect. 2.2.1); secondly, waste materials of consumption (Sect. 2.2.2); thirdly, several pollutants with centralized waste treatment

---

[2] For an "Introduction to Modern Austrian Capital Theory" see FABER (1979). Advanced studies in Modern Austrian Capital Theory are contained in FABER (1986) – For a summary of the distinctive features of our approach see FABER (1986:25–31, 45–48).

(Sect. 2.2.3); and fourthly, the differentiation between local and centralized waste treatment (Sect. 2.2.4).

2.1 The Basic Model

2.1.1 The Technology

We have represented the structure of production of the model of NIEMES (1981: Chap. 3) in Fig. 2.1. His model contains four production processes R_j ($j = 1, 2, 3, 4$) and five goods: labor, a capital good, a consumption good, an environmental good, and a pollutant. Each of the production processes takes one period. In R_1, l_1 units of labor are needed in order to produce one unit of the consumption good. Simultaneously s_1 units of pollutant (emission) are emitted within the same period:

$$R_1\colon l_1 \text{ units of labor} \rightarrow 1 \text{ unit of the consumption good} \oplus s_1 \text{ units of pollutant.}^3$$

The consumption good can also be produced in process R_2 which utilizes k_2 units of the capital good in addition to l_2 units of labor. The capital good here deteriorates with rate c, so that after k_2 units of the capital good have been utilized for one period only, $(1 - c)\,k_2$ units remain. We further assume that s_2 units of pollution (emission) are emitted from this production process:

$$\begin{aligned} R_2\colon l_2 \text{ labor} \oplus k_2 \text{ capital good} \rightarrow\ & 1 \text{ consumption good} \\ & \oplus (1 - c)\,k_2 \text{ capital good} \\ & \oplus s_2 \text{ emission.} \end{aligned}$$

The capital good is produced in process R_3, in which l_3 units of labor are needed per unit of the capital good. We assume for simplicity that there is no pollutant emitted by this process:

$$R_3\colon l_3 \text{ labor} \rightarrow 1 \text{ capital good.}$$

[3] These notations were taken from BLISS (1975:80).

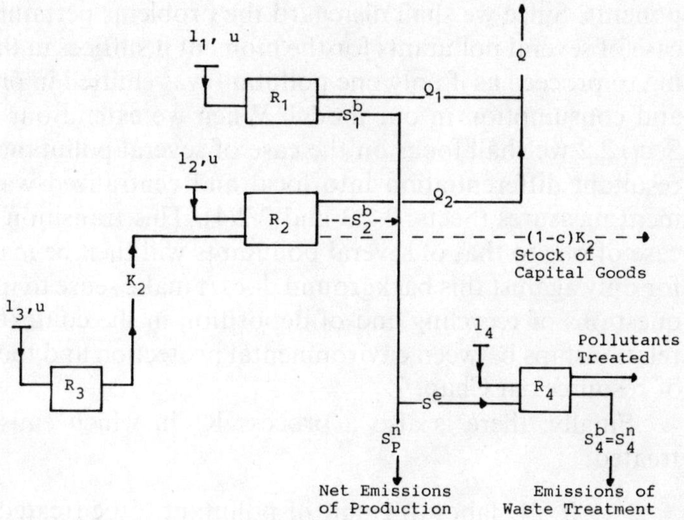

Q_j Quantity of Consumption Goods from Process R_j

Q Quantity of Consumption Goods

l_j Labor in Process R_j $(j=1,2,3,4)$

u Non-Scarce Environmental Goods

K_2 Quantity of the Capital Good Used in Process R_2

s_j^b Gross Emission of Process R_j $(j=1,2,4)$

s_P^n Net Emissions of Production

s^e Emissions to be Treated

c Rate of Depreciation

Fig. 2.1. Structure of the "Environment-Capital-Model" of NIEMES (1981: Chapter 3)

It is further assumed that in processes R_j $(j = 1, 2, 3)$ a flow of environmental goods containing a finite number of types of elements 1 to M is needed in addition to the factors labor and capital listed so far. This flow of environmental goods is not mentioned in the system of restriction equations because it is not assumed as scarce; this last assumption is of particular significance for later considerations (cf. Sects. 2.2.3 and 2.2.4 and Chap. 8 below).

It is important to note, however, that the case of several pollutants is a part of our model from the very beginning. On the emission side of the production processes R_j $(j = 1, 2)$ we are thus not only dealing with a pollution coefficient but with the respective vector of pollutants $s_j = (s_{1j}, \ldots, s_{Mj})$ with M pollution components. Since we shall disregard the problems pertaining to the case of several pollutants for the moment it suffices in the following to proceed as if only one pollutant was emitted in production and consumption in our model. When we extend our model in Sect. 2.2 we shall focus on the case of several pollutants and the resultant differentiation into local and centralized waste treatment measures (Sects. 2.2.3 and 2.2.4). This transition from the case of one to that of several pollutants will then be maintained; for only against this background does it make sense to investigate questions of recycling and of deposition in the context of interrelationships between environmental protection and the problem of resources in Chap. 7.

Finally, there is also a process R_4 in which emissions are treated:

$$R_4\colon l_4 \text{ labor} \oplus 1 \text{ unit of pollutant to be treated} \rightarrow 1 \text{ unit of pollutant treated} \oplus s_4 \text{ emission.}$$

In all these processes inputs are needed at the beginning of the period while outputs are available at the beginning of the following period. Pollutants, on the other hand, already arise during the period and are emitted and treated immediately. Note that it takes one period to produce the consumption good with process R_1 but two periods with R_2 and R_3, since it takes one period to manufacture the capital good first. If a resource is used to produce the capital good then it takes three periods to manufacture the consumption good (see Fig. 5.1). The structure of our model is represented in Fig. 2.1.

We now formally describe the temporal process of production and of waste treatment by constraint inequalities for each period t. For simplicity we assume that the quantity of labor L(t) in each period is constant. While not using these assumptions would make our derivations more complicated in parts, it would not essentially change the essence of our results. Since the production process R_1 requires labor only, this factor of production can always be fully employed.[4]

Let K_0 be the exogenously given quantity of the capital good at the beginning of period 1 and $x_j(t)$ the intensity with which

[4] This would not be true if R_1 also required the capital good. Cf. FABER (1979:98–104, 128) where the case of labor unemployment is analyzed.

process R_j (j = 1, 2, 3, 4) is run in period t. We then obtain for the labor constraint

$$L(t) - \sum_{j=1}^{4} l_j x_j(t) \geqq 0 \quad (t = 1, 2, \ldots) \tag{2.1}$$

and for the capital good constraint

$$K_0 + \sum_{g=1}^{t-1} x_3(g) - ck_2 \sum_{g=1}^{t-1} x_2(g) - k_2 x_2(t) \geqq 0$$
$$(t = 1, 2 \ldots). \tag{2.2}$$

The second term on the left hand side of (2.2) describes the sum of the investments made during the first (t − 1) periods; the third term summarizes the total deterioration during this time-span. The difference between these two terms is thus total net investment. If we add to this the initial stock K_0 we obtain the quantity of the capital good available at the beginning of period t.

The quantity of the consumption good Q (t + 1) available at the beginning of period t + 1 is produced in period t with processes R_1 and R_2:

$$Q(t+1) = \sum_{j=1}^{2} x_j(t) \quad (t = 1, 2, \ldots). \tag{2.3}$$

The quantity of gross emissions $S^b(t)$ is equal to the sum of the emissions of processes R_1 and R_2:

$$S^b(t) = \sum_{j=1}^{2} s_j x_j(t). \tag{2.4}$$

We assume for simplicity that the emissions of waste treatment $s_4 x_4(t)$ do not impede the quality of the environment. Since $x_4(t)$ units of the pollutant are treated in period t, net emissions $S^n(t)$ in period t are determined by

$$S^n(t) = \sum_{j=1}^{2} s_j x_j(t) - x_4(t) \quad (t = 1, 2, \ldots). \tag{2.5}$$

All variables $x_j(t)$, Q (t + 1), $S^b(t)$ and $S^n(t)$ (j = 1, 2, 3, 4; t = 1, 2, ...) must here be non-negative.

Let us take a look at Fig. 1.1 in order to put our model in its proper place. We see here that the following aspects are not yet taken into consideration in our approach: the emissions of consumption (arrow 4′), the utilization of capital goods in the waste treatment sector (part of arrow 2), and above all the utilization of resources (arrow 8) and thus, of course, recycling (arrow 2″). Before extending our model in these directions we think it useful to investigate the relationships between investment, the level of consumption, the abstention from a part of consumption, and the amelioration of the quality of the environment within this framework. We shall undertake this analysis in Sect. 2.1.3. First, however, we must describe the interdependence between the economic subsystem and the environment.

2.1.2 The Environmental Sector

Environmental problems are caused because the production of goods gives simultaneous rise to waste materials (emissions). Nature does have the ability to degrade the emissions (arrows 6, 6′, and 6″ in Fig. 1.1) discharged into the environment to a limited extent. Yet there remains a residue which as pollutants (arrow 7) affect the flow of environmental goods (arrow 8′) that reach the economic system (arrow 8). In the following we shall render the relationship between emissions, pollutants, and the flow of environmental goods through diffusion and damage functions.[5]

2.1.2.1 The Transformation of Emissions into Pollutants: The Diffusion Function

The net emissions $S^n(1)$ of the first period equal the difference between the sum of pollutants discharged in production in processes R_1 and R_2 on the one hand and the quantity of emissions treated in the first period on the other hand. A part of the latter was purified by nature within the first period. For simplicity we assume, as we did in (1.17), that this happens with a constant rate

[5] Cf. the diffusion function (1.9) of D'ARGE (1972) examined in Sect. 1.2.2.1 and the exposition of SIEBERT (1978:33).

of degeneration γ:

$$S^n(1) - \gamma S^n(1) = (1-\gamma) S^n(1) = \Phi S^n(1) \qquad (0 < \gamma \leqq 1, 0 < \Phi < 1); \quad (2.6)$$

the parameter Φ is called "diffusion factor". It shows what part of the emissions $S^n(1)$ remain in the form of pollutants, which are also called immissions, at the beginning of the second period:

$$D_1(S^n(1)) = \Phi S^n(1).$$

Likewise, the emissions discharged into nature during period 2, $S^n(2)$, result in $\Phi S^n(2)$ units of pollution in period 2. In order to be able to calculate the total quantity of pollution at the beginning of the third period we also have to take those pollutants $\Phi S^n(1)$ into consideration which were already present at the beginning of period 2. Assuming that during the second period they are also further reduced with rate Φ, we obtain from these a quantity of pollution $\Phi(\Phi S^n(1))$. Total pollution at the beginning of period 3 may then be calculated as the sum of the pollutant input of the respective periods, i.e.:

$$D_2(S^n(1), S^n(2)) = \Phi^2 S^n(1) + \Phi S^n(2) = \sum_{t=1}^{2} \Phi^{2+1-t} S^n(t); \quad (2.7)$$

D_2 is the diffusion function of period 2. Likewise, we obtain for period s:

$$\begin{aligned} S(s) &= D_s(S^n(1), \ldots, S^n(s)) \\ &= \sum_{t=1}^{s} \Phi^{s+1-t} S^n(t) \quad (s \in \{1, 2, \ldots\}). \end{aligned} \quad (2.8)$$

In the terminology used in Sect. 1.2 D_s denotes the quantity of pollutants S(t) which have accumulated in the environment over the whole time-span of t periods. The representation of the self-cleaning processes of the environment show clearly how important is the temporal structure of emission: If, with a given rate of degeneration γ a quantity of pollutants is so aportioned that it is evenly discharged into the environment over the course of an appropriate number of periods, the damage for nature and mankind is small. If, however, the total quantity is discharged within one or few periods, the damage is considerable.

2.1.2.2 *The Effects of Pollutants on the Flow of Environmental Goods: The Damage Function*

Damage functions indicate how the pollutant S(t) affects the flow of environmental goods U (t + 1) into the economic subsystem in period t + 1. SIEBERT (1978:33f.) for example utilizes damage functions with decreasing marginal damage in his economic models of the environment:

$$U(t+1) = G(S(t)) \geqq 0; \quad G'(S(t)) < 0, \quad G''(S(t)) < 0. \tag{2.9}$$

We shall now choose a specific type of this damage function. For this purpose we assume that U_0 is a flow of environmental goods unimpeded by immissions (arrow 8′ in Fig. 1.1) and that F (S (t)) is the environmental damage caused by the pollutant S (t). Instead of (2.9) we thus obtain (NIEMES 1981:62)

$$U(t+1) = U_0 - F(S(t)) \geqq 0; \quad F'(S(t)) > 0, \quad F''(S(t)) > 0. \tag{2.10}$$

The damage functions (1.9) of D'ARGE[6] (1972) and (1.18) of MÄLER (1974) discussed in the first Chapter are both functions of this type. Contrary to (2.10), however, these functions assume that marginal damage is constant instead of increasing.[7] Consider for example

$$U(t+1) = 1 - S(t)/V = 1 - D(t). \tag{1.18}$$

The concentration measure D (t) quantifies the burden of pollutants per unit of volume V in the environmental sector. It is linearly dependent on S (t). In order to simplify the presentation we shall in the following Sect. 2.1.3 use this specific damage function.

---

[6] If $U_0 = 0$ and if $U(t+1) = -S(t)$ is utilized, then (1.9) is a specific damage function with a negative sign.

[7] Besides increasing and constant marginal damage we may also encounter decreasing marginal damage. If the water from which drinking water is to be won is already heavily polluted, then it may well be the case that further pollution does not substantially increase costs if very costly purification processes are already in use anyway.

2.1.3 The Allocative Effects of Marginal Changes in the Production Program

We shall now analyze graphically the intertemporal effects of environmental protection measures taken within a period on the level of consumption and the quality of the environmental good in the succeeding periods. For this purpose we shall introduce small changes of production in each of the sectors. We shall demonstrate for instance, that a single reduction of consumption in one period only may improve the quality of the environment in many periods. Since economists often focus on the derivation of optimality conditions we would like to note that we are in this Section only indirectly concerned with such derivations. We shall rather demonstrate to the reader the complexity of the temporal structure of production. We believe that with our graphic analysis we shall contribute to a better understanding of the real relationships, and especially of the different temporal adaptation processes. Optimality conditions will be derived and interpreted in Sect. 2.2.1.2.

For our investigation we introduce the notion of a production program for the economic horizon of T periods. Such a production program is a vector π which lists the intensity with which each of the processes R_j ($j = 1, 2, 3, 4$) in period t ($t = 1, \dots, T$) is run. This vector thus contains 4T components.

Let π be a specific production program for the economic horizon of T periods. This production program is the *reference program*. We shall now change this program in period 1 as follows: The production of the consumption good in the first production process R_1 is decreased by a marginal unit. Since R_1 utilizes only labor as a factor of production, no other factor is released. We can use this quantity of labor in two ways. Firstly, we can increase the intensity of the waste treatment process R_4 in period 1. The reference program π thus changed is denoted by π_1. Secondly, we can use the quantity of labor to increase the intensity of the capital good process R_3 in period 1. This additional investment triggers a whole series of further changes with regard to the reference program π in the following periods. This new program is denoted by π_2. We shall now describe our procedure in detail.

1. *Production Program* π_1

During the first period we lower the production of the consumption good by one unit by reducing the amount of labor in process R_1 by l_1 units. At the same time emissions are thus reduced by s_1 units. The l_1 units of labor released are added in the waste treatment process R_4. In this way l_1/l_4 additional units of wastes may be treated in period 1 compared to the reference program π. The discharge of waste materials into the environmental sector in the first period is thus reduced by $(s_1 + l_1/l_4)$ units. As a result of the diffusion processes discussed in the preceding Section this lowers the rate of pollution at the beginning of the second period by $\Phi(s_1 + l_1/l_4)$ and thus, considering (1.15), leads to the following improvement of the quality of the environment

$$\Delta U(2) = 1 - \frac{S(1) - \Phi(s_1 + l_1/l_4)}{V} - \left(1 - \frac{S(1)}{V}\right)$$
$$= \frac{\Phi(s_1 + l_1/l_4)}{V}.$$

While consumption is reduced in one period only, the quality of the environment is improved in many periods; the degree of improvement, however, declines from period to period (cf. Fig. 2.2). The reason for this is that the $\Phi(s_1 + l_1/l_4)$ more units of waste material produced in the reference program is degraded in the course of time by natural processes. Thus the improvement of the environmental quality in period $t = 3$ only amounts to $\Phi^2(s_1 + l_1/l_4)/V$ and in period $t = s$ only to $\Phi^{s-1}(s_1 + l_1/l_4)/V$ units. As a consequence, the improvement of the environmental quality of production program π_1 compared with production program π diminishes over the course of time. Fig. 2.2 depicts both effects which are asymmetric with respect to time.

The shift of labor from process R_1 to the waste treatment process R_4 in period 1 causes a single decrease of the production of the consumption good, as opposed to a multi-period increase in the quality of the environment. This increase declines, however, over the course of time. The final reason for this temporal asymmetry is that consumption is a flow while the level of pollution is a stock. If the latter is decreased the quality of the environment is changed according to (1.18) as compared to the reference

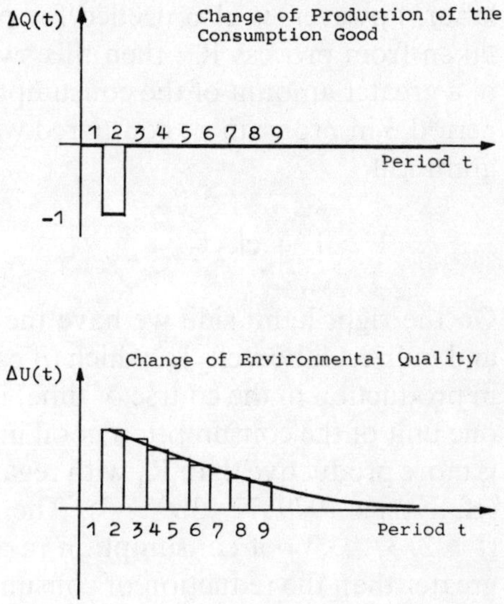

Fig. 2.2. Effects of a change of the intensity of production process R_1 in period 1

program π until the stock readjusts to the same level as that of the reference program. The change of the flow variable, on the contrary, only has a single-period effect.

With the help of Fig. 2.2 we can also explain why it is so difficult to make environmental policy measures politically acceptable. The reduction of consumption at the beginning of period 2 is directly and fully felt, while the improvement in environmental quality makes itself felt fully only in the course of time.

We note further that the lower the estimation of the value of future goods (i.e. the higher the social rate of discount δ) the lower is the weight that the improvement of the environmental quality carries in the decision about environmental policy measures.

2. *Production Program* π_2

The second possible program switch prescribes that the quantity of labor released, l_1, be utilized for an increase of the capital good investment in period 1. This leads to the additional production of l_1/l_3 units of the capital good in period 1. Here we can also observe a temporally asymmetrical effect, for investment changes a stock, in this case the capital stock.

From period P_2 onwards this investment of l_1/l_3 units of capital may be utilised by process R_2. A corresponding amount of labor, however, is also needed for process R_2. If this amount is taken from process R_1, then this switch leads to the availability of a greater amount of the consumption good at the beginning of period 3 in program π_2 compared with that in the reference program π if

$$l_1 > l_2 + ck_2 l_3. \tag{2.11}$$

On the right hand side we have the quantity of direct labor, l_2, and indirect labor, $ck_2 l_3$, which (if capital goods are fully used up in production in the course of time) is needed in order to produce one unit of the consumption good in R_2. If (2.11) is met, then R_2 is more productive than R_1 with regard to the consumption good[8] (cf. FABER 1979:72–3, 92–3). The sum of the increases $\Delta Q(t)$ $(t = 2, 3, \ldots, T)$ of consumption in all periods concerned is then greater than the reduction of consumption in period 1. As can be seen in Fig. 2.3 these increases diminish from period to period, as the quantity of capital goods added in process R_2 deteriorates at rate c.

The effects of this program switch for the quality of the environmental good are more complex than in the first case. On the one hand, this results from having to determine first the change in period 1 in which $x_1(1)$ decreases, only then can we determine the changes from period 2 onwards in which $x_2(t)$ is greater and $x_1(t)$ $(t \geqq 2)$ is smaller than in period 1. In addition we have to distinguish different cases because of the different relationship between emission and labor per unit of product in the production

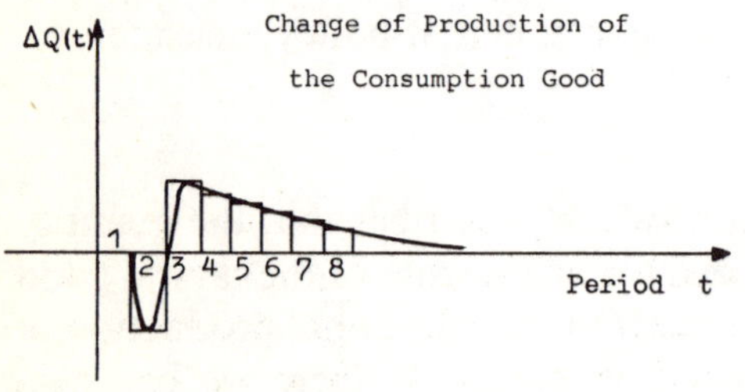

Fig. 2.3. Effects of an increase of investment in period 1 on the level of consumption

[8] The qualification "with regard to the consumption good" is necessary since the production with R_1 may possibly cause less ecologically damaging emission than the production with R_2.

processes R_1 and R_2. Let us begin with the effects in period 1 of a decrease of the production of the consumption good by one unit in process R_1; l_1 units are now added in process R_3. As a consequence the output of the capital good increases by l_1/l_3 units, while waste discharges decrease by s_1 units. The effects on the quality of the environment are thus similar to those in the production program π_1; the improvement, however, only amounts to

$$\Delta_{13} U(t) = \frac{\Phi^{t-1}(s_1)}{V} > 0 \quad (t \geqq 2).$$

We now turn to the effects that the program changes will have in periods $t \geqq 2$. As was mentioned above, the quantity of the capital good additionally produced in period 1 is used from period 2 onwards in process R_2. Since, if (2.11) is assumed, R_2 is more productive than process R_1, the amount of labor necessary for this is taken from R_1 and added to R_2. Thus from period 2 onwards R_2 is run with a higher intensity and R_1 with a lower.

What effects these changes have on the quality of the environment depends on the intensity of emission s_j/l_j in both processes R_j $(j = 1, 2)$. For this reason we define

$$\Delta_{jk} = s_j/l_j - s_k/l_k \quad (j, k = 1, 2; j \neq k).$$

If Δ_{12} is positive then the increase of the intensity of process R_2 and the decrease of the intensity of process R_1 in period 2 results in lower environmental pollution from period 2; the quality of the environment is improved. Contrary to the change of the quantity of emission in process R_1 caused in period 1 this type of change does not take place only once but for as long as the additional quantity of the capital good has not been fully used up. Since the capital good deteriorates at rate c, the quality of the environment is still improved in period 4, but this happens to a lesser degree than in period 3. The further development of the environmental quality depends on the values Δ_{12}, the rate of deterioration c, and the self-purification factor γ. We restrict ourselves here to this hint as a more precise demonstration of the stability of the system would require exact derivations (cf. NIEMES 1981:79–82). If Δ_{12} is negative, however, the development is reversed.

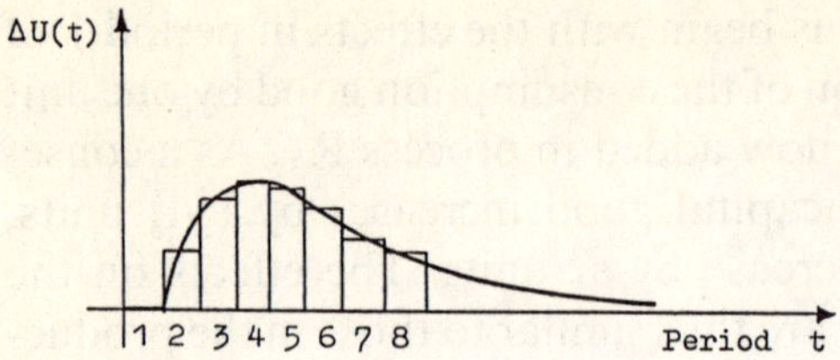

Fig. 2.4. Mutually reinforcing environmental effects of changes in production ($\Delta_{13} > 0$, $\Delta_{12} > 0$)

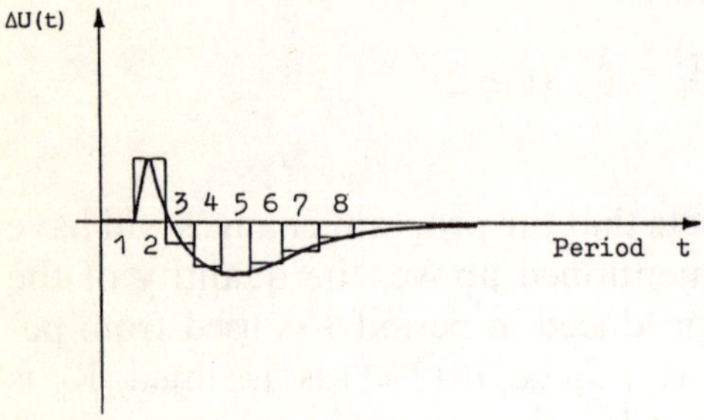

Fig. 2.5. Mutually compensating environmental effects of changes in production ($\Delta_{13} > 0$, $\Delta_{12} < 0$)

Let us summarize our analysis. Net emissions are changed only once, i.e. in period 1 as a result of the additional investment. This causes a multi-period improvement of the quality of the environment. The switch in period 2, i.e. the increased intensity of process R_2 and the reduction of process R_1, is superimposed upon this development. This causes a multi-period decrease ($\Delta_{12} > 0$) or increase ($\Delta_{12} < 0$) of the quantity of pollution from period 2 onwards, which results in an improvement or deterioration of environmental quality.

Two developments are depicted in Figs. 2.4 and 2.5. In Fig. 2.4 the effects of both types of program changes were assumed to reinforce each other ($\Delta_{13} > 0, \Delta_{12} > 0$). The case of compensating effects is depicted in Fig. 2.5.

Figures 2.4 and 2.5 can also be used to illustrate the significance of the length of the planning horizon (cf. NIEMES 1981: 88–90). If the case depicted in Fig. 2.5 is, for instance, restricted to a total of three periods, then a politician would support this change from an environmental policy point of view. If the planning horizon comprises eight periods, however, he would no longer support it, for the improvement of the environmental quality in period 3 is more than compensated by its deterioration in periods 4 to 8. In addition to the information supplied by Fig. 2.5, such a decision would of course require knowledge of the rate of time preference of the society.

2.2 Extending the Model

In the following we shall list five characteristics which, although important for environmental protection, are not embodied in the environment-capital model presented in the last Section.

1. Since labor is the only direct factor of production, the waste treatment sector in this model does not have a temporal structure of production. However, as was demonstrated at the beginning of Chap. 2, with the example of the BASF AG, the treatment of wastes can be very capital intensive. It is therefore important to take into consideration the application of capital goods in the waste treatment sector.
2. No distinction was drawn between local and centralized waste treatment, yet the local and subsidiary actions may in reality constitute a considerable part of such treatment; for the BASF AG, for example, it is more than 40% of total water protection expenditure.
3. We have so far allowed for one pollutant only. The emission of different pollutants, however, creates qualitatively different problems. Besides the consideration of synergetic effects in the environment[9], a suitable basis for the assessment for the harmfulness of emissions or pollution must be found.
4. Pollution attributable to consumption (cf. arrows 4′, 5′, and 6′ in Fig. 1.1) had not been taken into consideration.
5. The model did not allow for aspects of recycling and deposition, although these are closely linked to resource economics. As already mentioned in Chap. 1, it is these interrelationships between environmental protection and resource problems which are of particular importance.

[9] A synergetic effect arises when the combination of two chemicals creates an effect that is different from the sum of the two effects that are caused by each chemical on its own (VICTOR 1972:230).

This integration of the resource problems into our approach will constitute the object of Part III and particularly Part IV of this investigation. In this Section we shall, however, first show which extensions are necessary in order to be able to describe and incorporate the characteristics 1.–4. into our model. The use of a capital good in the waste treatment sector mentioned under 1, will here be dealt with more extensively than points 2.–4., for the thus extended model of Sect. 2.1 will, in Part IV, be combined with a resource model to be developed in Part III.

2.2.1 A Capital Good in the Waste Treatment Sector

We shall first describe the technology, then derive optimality conditions. Finally we shall interpret these extensively.

2.2.1.1 The Technology

We now take into consideration that not only in the production of the consumption good in process R_2, but also in the treatment of wastes in process R_4, the capital good is used as a factor input in addition to labor. For simplicity we assume that the capital good is homogeneous and can be used at will in either sector in each period. R_4 thus has the following form[10]:

R_4: l_4 units of labor $\oplus$ 1 unit of emission to be treated $\oplus$ k_4 units of the capital good $\rightarrow$ 1 unit of waste treated $\oplus$ $(1 - c_4)\,k_4$ units of the capital good $\oplus$ s_4 units of emission.

Instead of constraint (2.2) the constraint that is now obtained contains two additional summations

$$K_2(0) + K_4(0) + \sum_{g=1}^{t-1} x_3(g) - c_2 k_2 \sum_{g=1}^{t-1} x_2(g) - k_2 x_2(t) - c_4 k_4 \sum_{g=1}^{t-1} x_4(g) - k_4 x_4(t) \geqq 0 \quad (t = 1, 2, \ldots). \tag{2.12}$$

[10] Since throughout the rest of our investigation we shall refer exclusively to this waste treatment process it is unnecessary to use a different notation than that used in Sect. 2.1.1.

In addition to the distribution of the initial stock of the capital good

$$K(0) = K_2(0) + K_4(0) \qquad (2.13)$$

at time $t = 0$ among the processes R_2 and R_4, the two last terms have been added to (2.2). The first of these represents the quantity of the capital good which has been used up in the waste treatment sector up until time $t - 1$. The magnitude of this term is influenced by the rate of deterioration c_4, the capital coefficient k_4 and the levels of activity $x_4(g)$ $(g = 1, \ldots, t-1)$ of the waste treatment sector. The last term contains the quantity of the capital good $k_4 x_4(t)$ which is used in the waste treatment sector in period t.

All constraints on this model and the optimality conditions are given in the Appendix to this Chapter. As can be seen from a comparison of these constraint equations with those in Sect. 2.1, the "environment-capital model" is identical to our initial model except for the capital constraint (2.12). The essential characteristic of the model extension here undertaken is that in every period t a new decision can be made as to how much of the quantity of capital available is to be used in the production of the consumption good, and how much is to be used for the treatment of waste materials.[11]

In the following Section we shall derive and interpret shadow prices for factors of production and for the consumption good.

2.2.1.2 Optimality Conditions

We shall use the welfare function

$$W(Q(2), \ldots, Q(T), U(2), \ldots, U(T)) = \sum_{t=2}^{T} (1+\delta)^{1-t} W_t(Q(t), U(t)), \quad \delta > -1 \qquad (2.14)$$

as the intertemporal objective function. It has the usual properties

[11] For the sake of simplicity we employ only one capital good. If we had two capital goods this division would be irreversible.

(cf. KOOPMANS 1960), i.e. W_t is concave and especially

$$W'_Q(t), W'_U(t) > 0 \quad \text{and} \quad W''_Q(t), W''_U(t) < 0, \tag{2.15}$$

where δ is the rate of discount. The selection of this welfare function suggests itself for our investigation because it explicitly distinguishes between a temporal and an atemporal level of choice.

As already mentioned, the system of constraint equations of our model and the optimality conditions are included in the Appendix to this Chapter. From these we obtain the following relationships between prices and marginal utilities (cf. (i) and (ii) in the Appendix):

$$p_Q(s+1) = (1+\delta)^{-s} \partial W_{s+1}/\partial Q(s+1), \tag{2.16}$$

$$P_U(s+1) = (1+\delta)^{-s} \partial W_{s+1}/\partial U(s+1) \tag{2.17}$$

$$(s = 1, \ldots, T-1).$$

At the optimum the shadow prices of both dated goods are thus identical to their respective social marginal utility of the intertemporal welfare function. The social marginal utility is therefore determined by

a) the social discount rate δ,
b) the point of time s, and finally
c) the atemporal marginal utility of the consumption good, $\partial W_s/\partial Q(s)$, or of the environmental good, $\partial W_s/\partial U(s)$.

We shall now explain these optimality conditions in detail.

1. If the social discount rate is greater than 0 (i.e. if there is impatience to consume) then for constant marginal utility of the intertemporal welfare function the shadow price of the consumption and environmental goods decreases in the course of time. If, on the other hand, the social discount rate δ is smaller than 0, i.e. it lies on of the open interval (– 1,0) (patience to consume), then for constant marginal utilities the shadow price increases as t increases.
2. For processes R_1, R_2, and R_3 we obtain the following shadow prices for the production of one unit of the consumption or

capital good. These follow from the optimality conditions (iii), (iv), and (v) in the Appendix:

$$R_1:\ p_Q(s+1) = l_1\, p_L(s) + s_1\, p_{S^b}(s), \tag{2.18}$$

$$R_2:\ p_Q(s+1) = l_2\, p_L(s) + s_2\, p_{S^b}(s) + k_2\, p_K(s) + c_2\, k_2 \sum_{t=s+1}^{T} p_K(t), \tag{2.19}$$

$$R_3:\ \sum_{t=s+1}^{T} p_K(t) = l_3\, p_L(s). \tag{2.20}$$

If one thus produces one unit of the consumption good in R_1, one incurs (cf. (2.18)) $l_1\, p_L(s)$ labor cost[12] and $s_1\, p_{S^b}(s)$ cost as a result of gross emissions. In process R_2, on the other hand, costs for the capital good utilized are added (cf. (2.19)). $k_2\, p_K(s)$ costs arise for the services of k_2 units of the capital good and $c_2\, k_2 \sum_{t=s+1}^{T} p_K(t)$ for the deterioration of the capital good of $c_2\, k_2$ units. For the production of one unit of the capital good in R_3, $l_3\, p_L(s)$ labor costs are incurred (cf. (2.20)).

3. From the optimality conditions (vi), (vii), and (viii) in the Appendix we obtain for the shadow prices $p_{S^b}(s)$, $p_{S^n}(s)$, and $p_{S^e}(s)$ of gross and net emissions and of purified wastes the relationship

$$p_{S^b}(s) = p_{S^n}(s) = p_{S^e}(s) = l_4\, p_L(s) + k_4\, p_K(s) + c_4\, k_4 \sum_{t=s+1}^{T} p_K(t). \tag{2.21}$$

This indicates that the level of activity of waste treatment must be expanded until

(a) the marginal cost of waste treatment, $p_{S^e}(s)$, corresponds to the marginal damage of the net emissions in the environmental sector, $p_{S^n}(s)$, and

(b) the marginal cost for the neutralization of one further unit of gross emissions, $p_{S^b}(s)$, corresponds to the marginal damage that is otherwise caused in the environmental sector, $p_{S^n}(s)$.

---

[12] Due to the linearity of the technology marginal costs equal average costs.

The marginal cost of waste treatment consists of the marginal costs for the factors labor and capital used therein. As in process R_2, the marginal cost for the capital good is equal to the sum of the marginal costs of the quantity of the capital good used in the respective period s, $k_4\, p_K(s)$, and those marginal costs, $c_4\, k_4 \sum_{t=s+1}^{T} p_K(t)$, that arise from the fact that some uses in the following periods are foregone because the capital stock has deteriorated in the waste treatment sector.

For the shadow price of the capital good, that is, the sum of the services which can be obtained from one unit of the capital good, we use the notation

$$\bar{p}_K(s) = \sum_{t=s+1}^{T} p_K(t). \tag{2.22}$$

4. The marginal damage of the net emissions, $p_{S^n}(s)$, result from optimality condition (ix) in the Appendix:

$$p_{S^n}(s) = \sum_{t=s}^{T-1} p_S(t)\, \Phi^{t+1-s}. \tag{2.23}$$

What economic interpretation can we give to this term? As a result of diffusion there remain in period t only Φ units out of each unit of pollutant emitted during period t into the environment (cf. Sect. 2.1.2.1). Since only these Φ units burden the environment, the damage in period s is $p_S(s)\,\Phi$. In period s + 1 there are thus only Φ^2 units of the pollutant left, so that the environmental damage is only $p_S(s+1)\,\Phi^2$. If we add these to all other environmental damages for the whole planning horizon we obtain the price $p_{S^n}(s)$ for the marginal burden of net emissions, as indicated in (2.23). Only for the borderline case where net emissions in the environmental sector are always fully dissimilated during a period, i.e. the diffusion factor Φ is zero, do the emissions trigger no intertemporal external effects during the following periods.

5. In order to be able to compute total pollution damages we still need the shadow price of the pollutants. From the optimality

condition (x) in the Appendix it follows that

$$p_S(s) = -p_U(s+1)\,\partial G(S(s))/\partial S(s) \qquad (s = 1, \ldots, T-1). \quad (2.24)$$

At the optimum the shadow price of the pollution $p_S(s)$ is equal to the shadow price of the environmental good $p_U(s+1)$ weighted with the marginal damage $\partial G(S(s))/\partial S(s)$. We note especially that the shadow price of the pollution, $p_S(s)$, depends on S(s). It is thus influenced by the prior damage to the environmental sector.

6. If we replace the shadow price of the pollution in (2.23) by (2.24), and also consider Eq. (2.17), we obtain for the shadow price of net pollution

$$p_{S^n}(s) = \sum_{t=s}^{T-1} [(1+\delta)^{-t}\partial W_{t+1}/\partial U(t+1)] \, [(-G'_S(t))\,\Phi^{t+1-s}]. \quad (2.25)$$

The price of net emissions is thus determined by ecological factors (the diffusion factor Φ and the marginal damage $G'_S(s)$) on the one hand, and by economic factors (the social discount rate and the marginal utility of the environmental good) on the other hand.

2.2.2 Emissions in the Consumption Sector

As the next step we shall admit in our model emissions that arise in the consumption sector when quantities of the consumption good Q(t + 1) are consumed. Assuming for the sake of simplicity, as we did in the production sector, that in consumption we are dealing with a linear transformation process with the pollution coefficient s_C, in the consumption sector (C) we then have the following gross emissions:

$$S^b_C(t+1) = s_C Q(t+1) = s_C \sum_{j=1}^{2} x_j(t). \quad (2.26)$$

Table 2.1

Participant	Quantities of effluents TWA aver. (m^3/s)	TWA max. (m^3/s)	RWA max. (m^3/s)	Pollutant load (t BSB_5/d)
BASF + waste materials of the waste treatment plant itself	6.3	7.6	12.0	342.1
City of Ludwigshafen	0.8	0.8	1.1	25.0
City of Frankenthal + communes Bobenheim and Roxheim	0.5	0.5	0.9	7.9
Total	7.6	8.9	14.0	375.0

RWA, rainy weather discharge; TWA, dry weather discharge; BOB_5, biological oxygen demand; s, second.

One possible strategy to reduce these additional emissions is the construction of separate waste treatment installations for the consumption sector. Examples are communal sewage plants which purify mainly private effluents. Another possibility is the installation of a central waste treatment system, which caters for the emissions of the production sector as well as for those of the consumption sector. As can be seen from Table 2.1, the central effluent purification system of the BASF AG in Ludwigshafen/FRG also purified the effluents of the cities of Ludwigshafen and Frankenthal, and of other communes.

A combined effluent treatment system is advantageous as a rule, if only because the addition of communal effluents increases the content of biodegradable substances, which makes it much easier to purify industrial effluents in plants which work on a biological basis.

If the last mentioned waste treatment strategy, i.e. the centralized waste treatment, is what is wanted for the consumption sector, then the only change in our model is the extension of the constraint equation for gross emissions by the emissions from the consumption sector. Thus we obtain

$$H_{S^b}(s) = S^b(s) - \sum_{j=1}^{2} s_j x_j(s) - s_C \sum_{j=1}^{2} x_j(s-1) \geqq 0. \quad (2.27)$$

Without going into each of the steps of calculation, in analogy to our previous procedure for the model, extended by the gross emissions of consumption, we obtain the following shadow prices for the production of one unit of the consumption good:

$$\begin{aligned} R_1\colon\ p_Q(s+1) &= l_1 p_L(s) + s_1 p_{S^b}(s) + s_C p_{S^b}(s+1), \\ R_2\colon\ p_Q(s+1) &= l_2 p_L(s) + s_2 p_{S^b}(s) + s_C p_{S^b}(s+1) \\ &\quad + k_2 p_K(s) + c_2 k_2 \sum_{t=s+1}^{T} p_K(t). \end{aligned} \tag{2.28}$$

In both processes R_1 and R_2 the production costs are increased by the marginal cost of the emissions arising in consumption, $s_C p_{S^b}(s+1)$. Here we have to take into consideration the time-lag of one period between factor input and the disposal of the manufactured consumption good. Emissions from consumption therefore arise one period later than those in the production of the consumption good. The calculation of the marginal costs of the emission from consumption thus has to proceed on the basis of the shadow price $p_{S^b}(s+1)$. For the marginal costs of the emissions from production, however, the shadow price $p_{S^b}(s)$ of the previous period is valid.

If a joint treatment of the emissions which arise in production and consumption of the commodity is not aimed for, or if it is impossible for some other reasons, then separate waste treatment facilities have to be built up for the consumption sector. In principle there is little difference between this extended problem and the differentiation of the waste treatment sector into local and central waste treatment, which will be explained in Sect. 2.2.4. Before coming to this, however, we shall take a closer look at the case of several pollutants, as it is closely connected to the differentiation of local and central waste treatment measures.

2.2.3 The Case of Several Pollutants with Centralized Waste Treatment

2.2.3.1 The Assignment of Waste Treatment Costs

For simplicity, up to now it has been assumed that the same pollutant is emitted in each of the processes in the production sector. As was extensively discussed when we described the production processes[13], production requires environmental goods in addition to the factors of production labor and capital. Yet they were not considered in the system of constraint equations, since the flow of environmental goods into the production sector was not assumed to be scarce. As a result of the utilization of several environmental goods we thus get, as a rule, different emissions with differing harmful effects. The case of several pollutants was therefore, strictly speaking, contained in our model from the beginning. So far we have, however, not focused on the particular problems arising in this context. We shall now make good this omission.

The case of several pollutants raises the difficult problem of how a simple and practicable formula can be found with which treatment costs can be assigned to the different firms in accordance with the polluter-pays principle. For the assignment of waste treatment costs a measure is needed which specifies what waste treatment costs are caused per unit of each type of pollutant.

For practical purposes it is often necessary to summarize certain pollutants under a common denominator. For this purpose collective parameters are used, which should be as encompassing as possible. Examples are the BOD_5- and the COD- or TOC-value[14] in water quality management. The measure finally chosen is usually a compromise between the aims of ease of handling and sufficient exactness. It must also be taken into consideration that one single measure does not always suffice. For in addition to the cost of waste treatment, costs may also accrue for the remaining pollutants if they are charged for, as is the case for instance in the German Effluent Charge Act.[15] Since social and private objectives may be different, one must also differentiate between a measure devised from a social viewpoint and one devised from a

[13] Cf. Sect. 2.1.1.

[14] The TOC-value (total organic carbonite) measures the use of carbon dioxide (CO_2) when the effluent is burnt; the BOD_5-value on the other hand measures the biological and the COD-value the chemical oxygen requirements. For further parameters of water quality cf. ZWINTSCHER (1973:Chap. 2, Sect. 2.1).

[15] An excellent description and analysis of environmental, legal, economic, and empirical aspects of the German Effluent Charge Act and a comparison with the corresponding situation in the United States is given by BROWN and JOHNSON (1982).

private viewpoint. In the following we shall make this distinction in the context of our model.

2.2.3.2 Private and Social Aspects of the Assignment of Waste Treatment Costs and of Environmental Damage

Contrary to our procedure so far, S_p^b shall in the present and the following Section represent not only the gross emission of a waste material in production but instead a vector of M waste materials (see arrow 1 in Fig. 2.6). The same shall hold for the net emissions of production S_p^n (arrow 2), the waste materials to be treated centrally S^e (arrow 3), and the gross emissions of waste treatment S_4^b (arrow 4). As was explained above, the utilization of one measure really reduces the case of several pollutants to that of one pollutant. Thus if there are M different pollutants in an environmental medium such as water or air, the measure B is nothing but a projection from M-dimensional into 1-dimensional space. In the following we shall introduce two measures, one private and one social.

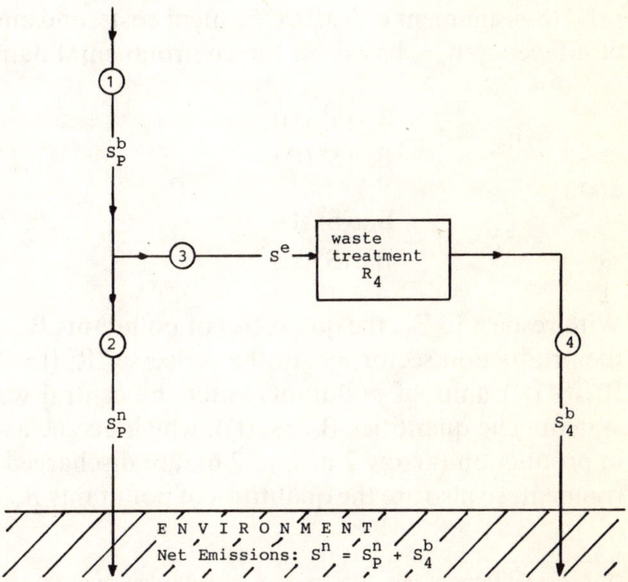

Fig. 2.6. Reassignment of waste treatment costs and environmental damage according to private and social criteria

We first turn to the private criterion. We assume that gross emissions of production S_P^b stem from several plants, a portion of them, S^e, is channeled into the waste treatment facility R_4 (see Fig. 2.6). For internal cost assignment purposes we now use a measure B_{S^e}. $B_{S^e}(S^e)$ thus indicates how many units of pollutants are contained in the vector S^e. In formulating such a measure, only those parameters which are representative of treatment costs will be used. The scope of the parameter chosen, the exactitude of measurement, and the resources spent for this measurement must be balanced.

Similar considerations apply to a measure which is to encompass the effects of the flow of emissions S_P^n and S_4^b discharged into the environment. The frame of reference for this measure, which we denote by B_{S^n}, is no longer the costs of waste treatment, but the actual environmental damage caused by these emissions.[16] These, however, are valued from a social viewpoint. It would thus be by pure coincidence that one and the same measure would suffice for both the distribution of waste treatment costs and the determination of environmental damage.[17] Therefore there are really two measures of effectiveness for a central waste treatment system; one measure of effectiveness $\eta_{B_{S^e}}$, which reflects the private assignment of waste treatment costs, and another measure of efficiency $\eta_{B_{S^n}}$, based on the environmental damage:

$$\eta_{B_{S^e}} = 1 - \frac{B_{S^e}(S_4^b(t))}{B_{S^e}(S^e(t))} \tag{2.29}$$

and

$$\eta_{B_{S^n}} = 1 - \frac{B_{S^n}(S_4^b(t))}{B_{S^n}(S^e(t))}. \tag{2.30}$$

With respect to B_{S^n} the quantities of pollutants $B_{S^n}(S_P^b(t))$ arise in the production sector, i.e. in the processes R_j ($j = 1, 2$), of these, $B_{S^n}(S^e(t))$ units of pollutants enter the central waste treatment system. The quantities $B_{S^n}(S_P^n(t))$, which accrue as net emissions in production (arrow 2 in Fig. 2.6), are discharged into the environment; so also are the quantities of pollutants $B_{S^n}(S_4^b(t))$, which

[16] A special form of this measure is indicated below in (2.36).

[17] The reason for this is that for the distribution of waste treatment costs the quantities actually emitted are used. Social damage is, however, caused by the pollutants.

accrue in the treatment of the wastes. Altogether gross emissions of production thus cause waste materials to the amount of

$$B_{S^n}(S^n(t)) = B_{S^n}(S^b_P(t)) - B_{S^n}(S^e(t)) + B_{S^n}(S^b_4(t)). \quad (2.31)$$

If we take into consideration that (2.30) yields for the quantities of emissions of waste treatment

$$B_{S^n}(S^b_4(t)) = (1 - \eta_{B_{S^n}}) B_{S^n}(S^e(t)), \quad (2.32)$$

then we can replace (2.31) by

$$B_{S^n}(S^n(t)) = B_{S^n}(S^b_P(t)) \left\{ 1 - \frac{B_{S^n}(S^e(t))}{B_{S^n}(S^b_P(t))} \eta_{B_{S^n}} \right\}. \quad (2.33)$$

The quotient $B_{S^n}(S^e(t))/B_{S^n}(S^b_P(t))$ indicates the portion of pollutants with regard to

$$\tau_{B_{S^n}} = \frac{B_{S^n}(S^e(t))}{B_{S^n}(S^b_P(t))} \quad (2.34)$$

which, out of all pollution arising in production is treated centrally. If we use (2.34) then (2.33) can be simplified as follows:

$$B_{S^n}(S^n(t)) = B_{S^n}(S^b_P(t)) \{1 - \tau_{B_{S^n}} \eta_{B_{S^n}}\}. \quad (2.35)$$

(2.35) indicates that the following actions lower the quantities of pollutants discharged into the environment with respect to the measure B_{S^n}:

1. A decrease in gross emissions arising in production,
2. an increase of that part of emissions which is centrally treated, and finally
3. an increase in the measure of effectiveness in the central waste treatment system.

In the following Sect. 2.2.4 we shall demonstrate that the three steps just mentioned can be effected by local waste treatment steps which is instituted in addition to central waste treatment.

2.2.4 Local and Central Waste Treatment Measures

Local waste treatment can be effected before or after central treatment. The purpose of such treatment can be, for instance, to change the vector of pollutant coefficients $s_j = (s_{1j}, \ldots, s_{Mj})$ of the j-th process so that the quantities of pollutants decrease with respect to either one of the measures mentioned above.

In order to demonstrate this we choose a certain measure B_{S^n}, namely

$$B_{S^n}(S^n) = \sum_{i=1}^{M} \alpha_i S_i^n . \tag{2.36}$$

For (2.35) we thus obtain the specification

$$B_{S^n}(S^n(t)) = (1 - \tau_{B_{S^n}} \eta_{B_{S^n}}) \sum_{i=1}^{M} \alpha_i S_i^b(t), \tag{2.37}$$

where $S_i^b(t)$ denotes the level of gross emissions for the pollutants $i = 1, \ldots, M$ at times $t = 1, \ldots, T$. For the two processes R_j $(j = 1, 2)$ of our model these can be obtained from the sum

$$S_i^b(t) = \sum_{j=1}^{2} S_{ij}^b = \sum_{j=1}^{2} s_{ij} x_j(t). \tag{2.38}$$

If this is inserted into (2.37) we obtain equation

$$B_{S^n}(S^n(t)) = (1 - \tau_{B_{S^e}} \eta_{B_{S^e}}) \sum_{i=1}^{M} \sum_{j=1}^{2} \alpha_i s_{ij} x_j(t). \tag{2.39}$$

Local treatment bring about a decrease in certain coefficients s_{ij} of the vectors s_j, and thus lower the level of waste discharges into the environment. Such local treatment will be enforced in that process j $(j = 1, 2)$ which causes most pollution. The treatment will concentrate on those waste components which are assigned high weights α_i within the pollution measure. Examples for such purposive local treatment is the belated installation of filtering systems and other physico-chemical procedures which can be inserted at the end of the production processes.[18]

It is of interest in this context that it does, after all, make a difference whether local steps are taken after the installation of a production plant, or whether environmental protection aspects are taken into consideration right from the very development of new production processes. Costs for parts that are installed later can be fairly accurately specified. Local waste treatment is applied either if the costs in the reduction of waste discharges are lower than in the central waste treatment system, or if special problems have to be solved. For example, heavy metals necessitate special measures in effluents because they disturb the biological Section of waste treatment installations, thereby reducing their effectiveness.

In new production processes, however, expenditure of environmental protection are, at least in part, integral to the research budget; environmental protection aspects are taken into consideration even before the development of the new process. As is generally the case in the assignment of production costs of joint products, it is difficult here to determine which costs were incurred for the product and which were incurred for the reduction of emissions. The actual costs for environmental protection are generally higher than is suggested by published figures.

Instead of applying local action to the production processes one can naturally also reduce the quantity of wastes emitted into the environment by increasing the effectiveness of the central waste treatment plant. The same applies if the part of the wastes that is channeled into the central plant is increased (cf. arrow 3 in Fig. 2.6). It is of particular importance, however, to subject only those emissions to central treatment which really do need to be treated. For instance, in effluent purification a separation of effluents that need to be treated from those that do not need treatment is reached through the installation of separating equipment (cf. FABER, NIEMES and STEPHAN 1983:27, 31). Thus not

[18] The following local measures were particularly successful at the BASF AG in Ludwigshafen/FRG with respect to relieving the central waste treatment system: 1. the removal of highly concentrated organic effluents, 2. the separate treatment of substances that were difficult to degrade biologically, 3. the reduction of the metal content, 4. the reduction of the discharge of anorganic acid, 5. the avoidance of the accumulation of neutral salts (cf. MALLE 1978:191 ff.).

only the size of the waste treatment plant needed, but also the charges can be diminished, which are paid according to the Effluent Charges Act for the residual content of the effluents that are discharged from the waste treatment system into the preflooder.

Our considerations show that local waste treatment measures can also be included in our model without very great difficulty. The waste treatment sector would have to be differentiated to include local waste treatment processes geared to particular pollutants from the different production processes.

The characteristics of environmental protection pointed out in the preceeding Sects. 2.2.1, 2.2.2, and 2.2.3, i.e. the use of capital in the waste treatment sector, the discharge of waste materials in consumption and the case of several pollutants, were also included into our approach as developed in Sect. 2.1.

The integration of the resource problem into our capital theoretical approach aimed at in Part III will prove to be considerably more complicated. Furthermore, this it is to be linked to the problem of environmental protection. For this it is necessary to give both problem areas a common foundation. To work out such a common foundation is the purpose of the following Part II.

Appendix: Constraints and Optimality Conditions for the Environmental Protection Model with Capital Formation in the Waste Treatment Sector

1. In our model in Sect. 2.2.1 the following constraints apply to the factors of production labor L and capital K, the consumption good Q, gross emissions S^b and waste materials treated S^e, net emissions S^n and pollution S, and finally the environmental good U:

$$H_L(s) = L(0) - \sum_{j=1}^{4} l_j x_j(s) \geqq 0,$$

$$\begin{aligned} H_K(s) = {} & K_2(0) + K_4(0) + \sum_{t=1}^{s-1} x_3(t) \\ & - c_2 k_2 \sum_{t=1}^{s-1} x_2(t) - k_2 x_2(s) \\ & - c_4 k_4 \sum_{t=1}^{s-1} x_4(t) - k_4 x_4(s) \geqq 0 \end{aligned}$$

$$H_Q(s+1) = \sum_{j=1}^{2} x_j(s) - Q(s+1) \geqq 0,$$

$$H_{S^b}(s) = S^b(s) - \sum_{j=1}^{2} s_j x_j(s) \geqq 0,$$

$$H_{S^e}(s) = x_4(s) - S^e(s) \geqq 0,$$

$$H_{S^n}(s) = S^n(s) - [S^b(s) - S^e(s)] \geqq 0,$$

$$H_S(s) = S(s) - \sum_{t=1}^{s} \Phi^{s+1-t} S^n(t) \geqq 0,$$

$$H_U(s+1) = G(S(s)) - U(s+1) \geqq 0,$$

$x_j(s)$, $S^b(s)$, $S^e(s)$, $S^n(s)$, $S(s) \geqq 0$ for $j = 1, 2, 3, 4$ and $s = 1, 2, \ldots, T$ and $Q(s+1)$, $U(s+1) \geqq 0$ for $s = 1, \ldots, T-1$.

2. The shadow prices for the temporally optimal use of the environment as a recipient of pollutants are obtained by optimizing the intertemporal objective function (2.14) under the above restrictions according to the procedure of KUHN and TUCKER (cf. e.g. HADLEY 1964: Chap. 6, 185–212). From the LAGRANGEAN function

$$V = \sum_{t=2}^{T} (1+\delta)^{1-t} W_t(Q(t), U(t)) + \sum_{t=1}^{T} \sum_{h} p_h(t) H_h(t),$$

in which L, K, Q, S^b, S^e, S^n, S, and U are replaced by h in order to simplify the notation, we obtain the following KUHN-TUCKER conditions for periods $s = 1, 2, \ldots, T-1$

(i) $$\partial V/\partial Q(s+1) = (1+\delta)^{-s} \partial W_{s+1}/\partial Q(s+1) - p_Q(s+1) \leqq 0,$$

(ii) $$\partial V/\partial U(s+1) = (1+\delta)^{-s} \partial W_{s+1}/\partial U(s+1) - p_U(s+1) \leqq 0,$$

(iii) $$\partial V/\partial x_1(s) = p_Q(s+1) - l_1 p_L(s) - s_1 p_{S^b}(s) \leqq 0,$$

(iv) $$\partial V/\partial x_2(s) = p_Q(s+1) - l_2 p_2(s) - s_2 p_{S^b}(s) - k_2 p_K(s) - c_2 k_2 \sum_{t=s+1}^{T} p_K(t) \leqq 0,$$

and for periods s = 1,...,T

$$\text{(v)} \quad \partial V/\partial x_3(s) = \sum_{t=s+1}^{T} p_K(t) - l_3 p_L(s) \leqq 0,$$

$$\text{(vi)} \quad \partial V/\partial x_4(s) = p_{S^e}(s) - l_4 p_L(s) - k_4 p_k(s) - c_4 k_4 \sum_{t=s+1}^{T} p_K(t) \leqq 0$$

$$\text{(vii)} \quad \partial V/\partial S^b(s) = p_{S^b}(s) - p_{S^n}(s) \leqq 0,$$

$$\text{(viii)} \quad \partial V/\partial S^e(s) = p_{S^n}(s) - p_{S^e}(s) \leqq 0,$$

$$\text{(ix)} \quad \partial V/\partial S^n(s) = p_{S^n}(s) - \sum_{t=s}^{T-1} p_S(t)\,\Phi^{t+1-s} \leqq 0,$$

$$\text{(x)} \quad \partial V/\partial S(s) = p_S(s) + p_U(s+1)\,\partial G(S(s))/\partial S(s) \leqq 0,$$

$$\text{(xi)} \quad \partial V/\partial h(s+1) \begin{cases} = 0 \text{ for } h(s+1) > 0 \\ \leqq 0 \text{ for } h(s+1) = 0 \end{cases}$$

$$h(s+1) = Q(s+1), U(s+1), (s = 1, \ldots, T-1),$$

$$\partial V/\partial x_i(s) \begin{cases} = 0 \text{ for } x(s) > 0 \\ \leqq 0 \text{ for } x_i(s) = 0 \quad (i = 1, 2, 3, 4), \end{cases}$$

$$\partial V/\partial h(s) \begin{cases} = 0 \text{ for } h(s) > 0 \\ \leqq 0 \text{ for } h(s) = 0 \end{cases}$$

$$h(s) = S^b(s), S^e(s), S^n(s), S(s) \qquad (s = 1, \ldots, T)$$

$$\partial V/\partial p_h(s) = H_h(s) \begin{cases} = 0 \text{ for } p_h(s) > 0 \\ > 0 \text{ for } p_h(s) = 0 \end{cases}$$

$$p_h(s+1) \geqq 0 \qquad \text{for } h = Q, U, S,$$

$$p_h(s) \geqq 0 \qquad \text{for } h = L, K, S^b, S^e, S^n \quad (s = 1, \ldots, t-1).$$

If in a period s all production processes are used and if all factors of production are also scarce then the equality sign is valid in the optimality conditions (i) to (xi).

Part II

Entropy and the Use of the Environment

> "'All flesh is grass', said the Prophet Isaiah. That is probably the most concise statement ever made of the ecological constraints on human life. But such vision of unity and wholeness has been fragmented by the specialization of modern thought. The economist's abstract world of commodities, with its laws of motion and equilibrium, has very few points of contact left with 'grass' and even is in danger of losing touch with 'flesh'. Seemingly, economics has become detached from its own biophysical foundations." (DALY 1973:33)

As has been pointed out in the first Chapter[1], there exist close mutual relationships between the different forms of use of the environment and thus also between resource and environmental problems. At this point we shall demonstrate this once again with three examples: one concerns the use of nuclear power, another the extraction of gravel in the Rhine Valley, and a third the treatment of effluents.

Example 1. Although nuclear power is widely used for providing energy there is to date no satisfactory answer to the question whether the reprocessing or the entombment of spent elements from conventional nuclear power plants should be adopted as a long range waste treatment strategy. Security considerations play a role in the solution of this problem as well as the growing scarcity of natural deposits. In order to avoid having to switch off nuclear power plants as a result of the looming waste treatment bottleneck, attempts are being made to win time with temporary solutions such as the compact or intermediate storage of spent fuel rods until one can establish a waste treatment strategy which satisfies the need both for environmental protection and for a secure energy supply.[2]

Example 2. The extraction of gravel in the Rhine Valley leaves flooded gravel pits which after costly reclamation[3] can be used above all for recreational purposes. Before this happens, however, valuable swamps, pastures, or reed landscapes are destroyed, which formerly served as a habitat for rare water birds

---

[1] Cf. especially Section 1.2.

[2] Even the contribution of I. C. BUPP to the "Energy Report" of the Harvard Business School (1979:150–191) takes a sceptical stance towards the question whether, in the face of the to date unsolved waste treatment problems, nuclear energy can in the long-run play a central role in the securing of the energy supply.

[3] Cf. BUCHWALD and ENGELHARD (1980, III:Chap. 6.4).

or for other endangered species. At the same time the extraction of drinking water is impaired, for the mining of successive layers of gravel weakens the natural purification of surface waters; it also reduces the groundwater area because the rate of evaporation is much greater over open waters than it is in swamp areas.[4]

Example 3. The purification of effluents in fully biological purification plants protects our waters from being further polluted. Yet it also burdens our environment in other ways, for the purification of industrial effluents and the burning of the accumulated sewage sludge is often very energy and raw material intensive. For instance, the main purification plant of the BASF AG in Ludwigshafen on Rhine, which has a purification capacity of 6.5 million "inhabitant equivalents"[5], needs approximately 22,000 kilowatt[6] electrical current per daily quantity of effluents of approximately 450,000 m^3. This equals the energy consumption of a medium-sized town like Frankenthal with a population of circa 50,000. Not only the drastic rise of energy prices in the last years makes it thus understandable that since the plant has been put into service at the end of 1974, efforts to curb energy consumption have been increased.[7] In the combustion of sewage sludge,

---

[4] In its third special report on "Environmental Problems of the Rhine" the Council of Experts for Environmental Problems notes e.g.: "At the Lower Rhine one estimates the supply of groundwater to decrease by 200,000 to 250,000 cubic meters per annum per km^2 of such areas. The gravel industry at the Lower Rhine estimates the area needed annually for the mining of gravel at around 15 square kilometers provided that the rate of output remains constant. This would amount to an area of 375 square kilometers until the end of the century. If these figures remain valid, the annual supply of groundwater at the Lower Rhine will decrease by the rate of output of a medium-sized waterworks in the future."

[5] One "inhabitant equivalent" corresponds approximately to a BOD_5-content of 60 g per day at 60 l.

[6] See e.g. ENGELMANN (1976) or ENGELHARDT and HALTRICH (1978) for more exact values.

[7] The technical control attempts to lower energy consumption by switching the aeration gyros in the activation basin on and off, which are described in BENDER and KONRAD (1977), were soon successful. In ENGELHARDT and HALTRICH (1978:515) one can read: "In the trial runs so far the regulation of the oxygen concentration lowered the expenditure for degeneration from an average of 1.1 kWh/kg degenerated BOD_5 to 0.77 degenerated BOD_5 for a 95% degeneration of the biochemical oxygen need." (Our translation)

energy was saved by adding coal dust instead of heavy oil to the burning process. Further decreases of cost in the combustion of sewage sludge are expected from current attempts to burn the sewage sludge in the firm's own power plant. These attempts are important because bottlenecks are already being foreseen for the medium run in the disposal of those waste materials, and sewage sludge which due to its high pollutant content, cannot be used as fertilizer on agricultural lands.[8] The existing depots will soon no longer suffice to receive all sewage sludge, the accrual of which has increased as a consequence of more rigorous water protection. New depots, however, cannot be developed in the short run. For the road from planning, via approval, to the actual installation of a depot is lengthy. To summarize: water protection, i.e. the reduction of the use of the environmental medium water as a recipient of pollutants, leads to an increase in raw material and energy consumption and, furthermore, to a shift of the environmental problems on to other environmental media, namely air and ground.

These three examples show that the question of choice of a waste treatment strategy which is both secure and economically viable in the long-run, is closely connected to given stocks of resources and their development.[9] The resource problem thus has to be taken into consideration in an analysis of alternative waste treatment strategies. In the first two Chapters, however, it was only touched upon lightly, since our focal point there was the function of the environment as a recipient of pollutants.

Before investigating the interrelationship between environmental protection and resource problems, however, we have to accomplish several preliminary tasks. In Chap. 3 we shall first introduce the notion of entropy. This notion enables us to put the uses of the environment as a recipient of pollutants and as a

---

[8] These bottlenecks in the deposition of sewage sludge and of the Neckar excavation volume were pointed out several times at the 1. Neckar-Environment-Symposium on Feb. 16th and 17th, 1981 in Heidelberg/FRG, which was sponsored by the Baden-Württemberg Ministery of Nutrition, Agriculture, Environment, and Forestry and the Institute for Sediment Research of the University of Heidelberg.

[9] In a long range decision as to the form of waste treatment the actual choice is, of course, very much influenced by the technology and the range of products; these in turn are influenced by waste treatment possibilities.

supplier of resources on a common basis. In Sect. 3.8 we shall show that the ecological equilibrium situation can be characterized by a constant level of entropy of the environmental sector. Disturbances of the environmental sector resulting from emissions of wastes from the economic system can be described by an increase in entropy. In Sect. 3.9 we present a first approach to the problem of how the notion of entropy can be introduced into economic analysis. In Chap. 4 we shall demonstrate that the extraction of resources is equivalent to an exploitation of a volume of ecological services, which can be equally interpreted as a change in entropy. Our starting point is here the interpretation of the separation process as the reversal of a diffusion process. The extraction of resources can thus be interpreted as the reversal of a diffusion process. Diffusion processes are, as a rule, irreversible. The example of the process of heat expansion vividly describes the diffusion process. For this reason the utilization of the tools of thermodynamics suggests itself. For, on the one hand it enables us to derive relationships between entropy, the stock of resources, and the concentration of resources in the environmental sector. On the other, it allows us to establish relationships between the concentration of resources and energy requirements in resource extraction.

We would like to close this introduction by providing the reader with a first hint as to how the notion of entropy can be used for economic considerations. The following quote from *Global 2000* seems to the point:

> "Entropy is the scientific term that measures how 'mixed up' things are. 'Concentratedness', as used above, is the opposite of entropy. Concentrated deposits have relatively low entropy and low production costs; low grade deposits have high entropy and high costs. Combining minerals and materials in alloys and other products and distributing these products throughout the economy increases the entropy of the minerals used; how much entropy is added determines largely the practicality of recycling. The factors that determine the entropy of ores and the wastes from society are therefore highly relevant for maintaining an adequate stock of minerals in use at manageable costs." (Global 2000 1981:220)

As we shall see, the notion of entropy can be used to analyze the function of the environment as a supplier of resources. We shall show that this is also true for the use of the environment as a recipient of pollutants.

3 The Notion of Entropy[10]

> "But – a fact hard to explain – loud though the noise caused by the Entropy Law has been in physics and the philosophy of science, economists have failed to pay attention to this law, the most economic of all physical laws." (GEORGESCU-ROEGEN 1971 : 280)

> "If we will be more forthcoming with explanations of our cherished terms, our science colleagues may be more enclined to help us out with 'entropy', which to me is a more difficult concept than anything economics has to offer." (KOOPMANS 1979:13)

3.1 Thermodynamics

Thermodynamics is that discipline of physics which is concerned with the behavior of systems that contain a large number of particles. Thermodynamics involves two important alterations if compared to classical mechanics, that part of physics which is concerned with the movements of a small number of particles. These are explained in the following:

1. Classical mechanics describes systems which contain few particles only. The planetary system can be considered as a typical example. Here it has to be taken into consideration that, on a closer look, this system really consists of an incredibly large number of molecules. We are only interested in the movement of the planets, however, so that each planet is considered to be a single particle. Such a planetary system can be fully described by specifying mass, position, and velocity of each planet in-

[10] We acknowledge the coopertation of Dipl.-Phys. Lutz FREYTAG in Chaps. 3 and 4.

cluding its central star (the sun). Knowledge of these magnitudes allows us to calculate all further movements of this planetary system.

Let us consider the air contained in a balloon, on the other hand. The air is known to consist of different molecules which collide similar to billiard balls. Each of these collisions obeys the laws of classical mechanics. If one knew the position, velocity and mass of each of these molecules as one does in the planetary system, one could calculate the movement of the molecules in a balloon, too. There are two reasons, however, why this is impossible in practice.

For one thing a balloon containing one mol of gas contains approximately 10^{23} molecules. Due to this incredibly great number it is not feasible to determine all positions, velocities, and masses; even if this was achieved the further behavior of the system could not be calculated because the problem would be too complex for even the largest computer.

Another thing is that the information about the movement of each molecule in the balloon would be without any practical use. Rather, we are much more interested in measuring the balloon's volume, its pressure, its temperature. These are magnitudes, however, which play no role in the description of systems in classical mechanics. Here it suffices to know positions, velocities, and masses – as was seen in the example of the planetary system.

Let us summarize as follows. The description of systems comprising a large number of particles – in the following we shall call these systems "thermodynamic systems" – makes it necessary to introduce entirely new physical magnitudes. Thus it is the task of thermodynamics to define useful magnitudes and to determine the laws that these magnitudes follow.

Some of these terms that have to be newly defined, such as temperature or pressure, are familiar to us from everyday life. One, however, the notion of entropy, plays no part in everyday language. It is the aim of the present Chapter to introduce this notion and to describe the laws that are connected with it.

2. Let us return to the planetary system. It is an important result of classical mechanics that all processes described by it are temporally reversible. For the planetary system this means that celestial bodies could also move in the opposite direction

– i.e. as if a film in which their movements are recorded was run in reverse. Movements in both directions are equally valid in principle; which of them is realized depends only on the primal movement of time. The following fact, however, is of importance; were our planets slowed down and then set back in motion in the opposite direction, then this movement would be fully identical to that of the film running in reverse. Such a process which can be "turned around in time" is called "reversible".

A totally different situation is encountered in a thermodynamic system. Let us for instance consider an iron rod which is being heated at one end. After the heat input is finished the temperature will equalize throughout the rod; this will occur according to the Law of Fourier (1811). Yet one will never observe the opposite process in nature. Within an iron rod, throughout which temperature is at an even level, heat will never spontaneously, i.e. without any outside influence, accumulate at one end while the other cools down. A process of the same kind as that of the diffusion of heat through an iron rod is called "irreversible".

Apparently we need an additional law of nature in order to describe thermodynamic systems. This law is to characterize the direction of the processes involved. We shall see that such a law can be very elegantly formulated with the help of the notion of entropy.

3.2 Energy, Gibbs' Fundamental Equation, Intensive and Extensive Quantities

The physicist attempts to describe nature by using notions which are, above all, physical quantities. One of the most important physical quantities is energy. With the help of this notion it is possible to establish a relationship between phenomena as different as the attraction of two electrically charged particles, a chemical combustion process, and the heating of a physical body:

> "The phenomena which are summarized by the notion of energy are so numerous and so dissimilar, however, that it is not possible simply to enumerate all of them. They concern all processes which we observe in

> nature. Energy is a tie, which allows us to recognize the dependencies between *all* appearences of nature. The variable energy is here so constructed that these dependencies express themselves in the *preservation* of energy: *energy can be neither created nor destroyed.*" (FALK and RUPPEL 1976:1, their emphases; our translation)

This is the First Law of thermodynamics. It is the task of thermodynamics to describe the changes of state of systems. Such changes of state are described by changes in physical quantities. In order to obtain a simple description of such changes of states it has proved useful to consider first of all infinitely small changes, i.e. (in economic terminology) marginal changes. It becomes clear from what was said above that changes of energy are of particular importance in such a system.

At this point the notion of a change of energy in a system needs more specific explanation. It is true that, according to the quote from FALK and RUPPEL, energy can be neither created nor destroyed; yet the energy of a particular subsystem can change. Let us consider for example a freshly filled fireplace. After the wood has been burnt, the energy available in the form of chemically bound energy has been transformed into a higher room temperature. The energy content of the fireplace has thus been reduced by the same quantity by which the energy of the air has increased. At the same time, however, a transformation of energy from one form into another, from chemical to heat energy, has taken place.

Let us consider another example. Consider a cylinder filled with air be closed by a piston. The piston is to be moved inwards by a small distance dx. What amount of energy is needed? As is well known, the mechanical form of energy is called work, which is defined as the product of force and distance dx. If the force, which the air exerts on the piston, is denoted by F and the infinitely small change of energy by dU, one obtains

$$dU = F\,dx.$$

As was explained in the previous Section, thermodynamics generally uses quantities such as pressure and volume. It is now fairly simple to transcribe the law just obtained to these quantities. Let the area of the piston be of size K; considering that the volume of a cylinder is given by the product of area and height, the definition of pressure p as force divided by area yields

$$dU = \frac{F}{K} K\,dx = (-p)\,dV. \tag{3.1}$$

The minus-sign in this equation is to take into consideration that the change of volume dV refers to a reduction in the volume of the gas. The sign signifies that a decrease of volume $dV < 0$ goes along with a positive change of energy. The compression has increased the energy of the air in the cylinder.

This new form of the equation has a considerable advantage if compared to the old one. For it is no longer bound to the specific geometry of the change of volume considered by us, but depends only on the pressure and on the change of volume.

From the equation yielded we can see how useful it is to restrict a first analysis to infinitely small changes of state. Had we considered, for instance, a large change in the position of the piston, we would have had to take into consideration that the pressure also changes with increasing compression (i.e. with diminishing V). If such a law is known, however, as in the case of the universal gas equation

$$pV = NRT, \tag{3.2}$$

where p denotes pressure, V volume, N the number of mols, R the universal gas constant, and T the absolute temperature (counted in degrees Kelvin)[11], the change of energy can – for a finite change of volume – now be very easily specified.[12] Let us consider the case of constant temperature in this context. In reality this means that we press the piston only very slowly into the cylinder;

[11] The introduction of a Kelvin scale for the temperature represents a considerable simplification from the physical point of view. Thus the universal law of gases would have to be written in the form

$$pV = NR\,(T - 273.15°)$$

if the Celsius scale was used. The Kelvin scale is of particular significance also because a temperature of − 273.15 °C, i.e. 0 °K, is the absolutely lowest temperature obtainable. The temperature which is measured with the help of the Kelvin scale is thus called absolute temperature, for it starts measuring at the absolute zero point of temperature.

[12] For the special case of constant temperature the universal gas law reduces to the BOYLE-MARIOTT-LAW $pV = \text{const.}$

for otherwise the gas would warm up just as it does in a bicycle pump which is operated quickly. Due to (3.2) we then get

$$p(V) = \frac{NRT}{V} \tag{3.3}$$

and thus, if the volume of gas is changed from V_1 to V_2, for the change of energy due to (3.1) and (3.3):

$$U = -\int_{V_1}^{V_2} p(V)\,dV = \int_{V_2}^{V_1} p(V)\,dV = NRT\,(\ln V)\big|_{V_2}^{V_1}$$

$$U = NRT \ln \frac{V_1}{V_2}. \tag{3.4}$$

Laws similar to the form derived above, for an infinitely small change of energy when there is a mechanical change of state $dU = -p\,dV$, can now be derived for any change of state. This is, for instance, easily possible for changes of electrical, chemical, or kinetic changes of states. Among all these formulae there is an important similarity. The infinitely small change of energy can always be expressed as the product of a physical quantity with an infinitely small change (marginal) of another physical quantity:

$$dU = (\text{phys. quantity}) \cdot d\,(\text{phys. quantity}). \tag{3.5}$$

Both physical quantities have to be given, of course, in such units that their product yields units of energy.

It cannot be decided, however, from this circumstance alone, whether $p\,dV$ or $V\,dp$ is the right formula for the change of energy accompanying mechanical changes of states. Yet here, too, an observation helps. For one may take significant notice of the fact that physical quantities, whose product yields units of energy, may be classed in two groups. This classification results from the different behavior of these quantities when two similar systems are being coupled. One differentiates between intensive quantities, i.e. quantities which take on identical values in both systems when these are coupled; and extensive quantities, whose value for the total system is simply the sum of the values of these quantities in both systems. Let us take as an example two closed, gas-filled cylinders which are mutually connected by a pipe. This connection causes the pressure in both cylinders to be always identical. Pressure is thus an intensive quantity.[13] The volume of the total system is simply a result of the sum of the volumes of both subsystems. Thus the volume is an extensive quantity of the system. This result can be generalized as well:

$$dU = \text{intensive quantity} \cdot d\,(\text{extensive quantity}). \qquad (3.6)$$

The infinitely small change of energy of a system is equal to the product of an intensive quantity with an infinitely small change of an extensive quantity. Another example of an extensive quantity is the number of particles of a system. The corresponding intensive quantity is the chemical potential, an often used quantity in chemistry.

With the help of these results it is possible to determine a general formula for the infinitely small change of energy:

$$dU = \sum_i X_i \, dY_i , \qquad (3.7)$$

where X_i denote the intensive and Y_i the extensive quantities. This formula is called GIBBS' Fundamental Equation. It is the basis of thermodynamics.

3.3 An Example of Entropy: The Diffusion of Gases

In the previous Section we have seen GIBBS' Fundamental Equation, which describes the changes of the internal energy of a system with respect to every form of energy. Now we pose the question of which quantities can be used to describe the heat form of energy. For this we need an intensive quantity and an infinitely small change of an extensive quantity, whose product yields the change of this form of energy.

---

[13] This very effect is used in hydraulic presses in which the cylinders are closed by pistons of different sizes; due to (3.1) and $F = pK$ one thus obtains different forces on the pistons.

The physical quantity that is most important in the context of heat phenomena is temperature. In order to decide whether temperature is an extensive or an intensive quantity we consider two pieces of iron of different temperature. If these two pieces are brought into contact with each other so that the heat form of energy contained in both can flow from one to the other – this is usually the case if both pieces of metal simply touch – the temperature in both pieces of metal will equalize. Thus we recognize that temperature is the intensive quantity to the heat form of energy.

It is considerably more difficult to find a suitable extensive quantity which, if multiplied with temperature, yields units of energy. An extensive quantity with this property was originally unknown. Therefore CLAUSIUS (1822–1888) simply defined a new physical quantity with exactly these properties. GIBBS' Fundamental Equation can be interpreted as the formula that defines this new quantity:

$$dU = TdS - pdV. \tag{3.8}$$

We have here abbreviated this new quantity by the letter S. It is called entropy. We expect the quantity entropy to have an extensive character, so that the entropies of two subsystems can simply be added when these subsystems are joined to form one system.

Before we proceed we note that (3.8) is a relationship from equilibrium thermodynamics. This implies that all corresponding processes take place at an infinitely slow speed. For all realistic economic processes (i.e. for all non-isolated systems) we obtain for the change in heat

$$TdS = dU - |\partial A|,$$

where ∂A is the work which has been performed on the system.[14] In all these cases, however, (3.8) can be used to find a lower bound for dS. Since it is rather cumbersome to work with inequalities we shall use equalities. This implies that our analysis of economic and ecological problems is correspondingly restricted (FABER 1985:321–2).

[14] ∂A is not a state variable but $\int_a^b \partial A$ depends on the corresponding physical process.

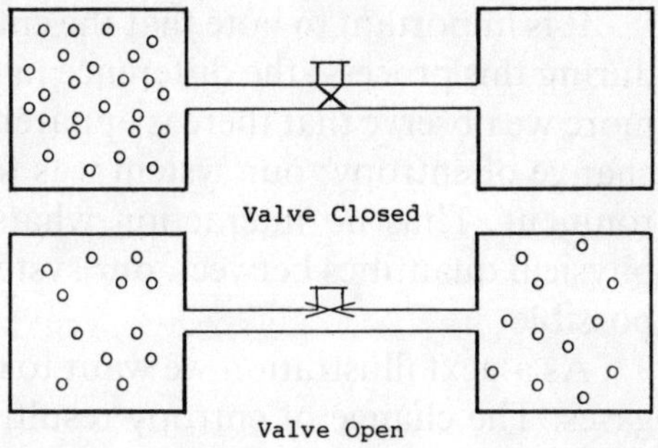

Fig. 3.1. Expansion of a gas

In the following we want to show how the change of entropy of a gas can be calculated. For this purpose we consider a container filled with gas which is connected to another evacuated container via a closed valve. Let the volume of the filled container be V_1, and that of the evacuated container V_2. Let the two containers also be so isolated as to allow no exchange of energy with the environment. If the valve is now opened the gas flows from one container into the other (cf. Fig. 3.1). Let us assume that the gas obeys the universal gas law $pV = NRT$. Let us also take into consideration that due to the isolation there is no exchange of energy with the environment, so that the energy of the gas remains constant. As a result of (3.8) we then obtain

$$0 = dU = T\,dS - p\,dV = T\,dS - \frac{NRT}{V}\,dV$$

and thence

$$dS = \frac{NR}{V}\,dV.$$

For the change of entropy between the two states we obtain by integration

$$\begin{aligned} S(V_1 + V_2) - S(V_1) &= NR \int_{V_1}^{(V_1+V_2)} \frac{1}{V}\,dV \\ &= NR \ln\left(\frac{V_1 + V_2}{V_1}\right) > 0. \qquad (3.9) \end{aligned}$$

It is important to note that the entropy of the system increases during this process; the difference in entropy is positive. Furthermore we observe that there is apparently nothing that created this change of entropy; our system was totally isolated from its environment. Thus no interaction whatsoever, no exchange of any physical quantities between our system and its environment, was possible.

As a next illustration we want to describe the diffusion of two gases. The change of entropy resulting from this process can be easily obtained from the foregoing example. Before proceeding we note that in the following we use the mixture of ideal gases as an idealized didactical example for the qualitative explanation of the changes of entropy which occur if changes of concentration take place.

Let us take another look at the two containers, assuming this time, however, that the left hand container is filled with a gas of type 1 while the right hand container is filled with a gas of type 2 (cf. Fig. 3.2). An important insight of gas theory is that these two gases do not interact in any way once the valve has been opened. Thus we shall only observe that a part of the gas of type 1 flows from volume V_1 to volume V_2, while a part of the gas of type 2 flows from volume V_2 to volume V_1.

Since entropy is an extensive quantity, the entropy change yielded by this process is nothing but the sum of the changes in entropy of each of the two systems. These are determined, however, by formula (3.9) indicated above. Thus we obtain

$$\begin{aligned} \Delta S_1 &= N_1 R \ln\left(\frac{V_1 + V_2}{V_1}\right), \\ \Delta S_2 &= N_2 R \ln\left(\frac{V_1 + V_2}{V_2}\right), \\ \Delta S = \Delta S_1 + \Delta S_2 &= R\left(N_1 \ln\left(\frac{V}{V_1}\right) + N_2 \ln\left(\frac{V}{V_2}\right)\right); \end{aligned} \tag{3.10}$$

in these equations the number of mols of the gas of types 1 and 2 is denoted by N_1 and N_2. For simplicity we also used the notation $V = V_1 + V_2$ to indicate the total volume of the system. This equation will turn out to be very important for our study.

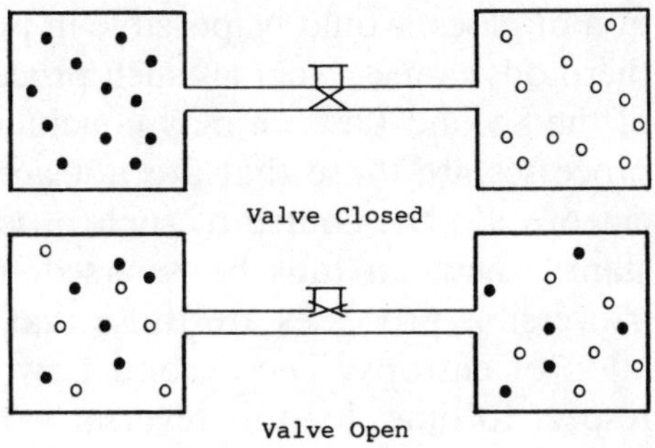

Fig. 3.2. Diffusion of two gases

3.4 The Second Law of Thermodynamics

So far we have succeeded in defining appropriate physical quantities to describe thermodynamic systems. In GIBBS' Fundamental Equation we have also found an important law that connects these physical quantities with each other. As was explained in Sect. 3.1 above, it is important to postulate a law that indicates in which direction observable thermodynamic processes evolve. As was also mentioned in Sect. 3.1 this law can be especially elegantly formulated with the help of the notion of entropy.

The processes described in the foregoing Section, i.e. the expansion of a gas into a vacuum and the diffusion of two gases, are typical irreversible processes. These processes only evolve spontaneously but in one direction. A gas which is evenly distributed within the two containers will never spontaneously retreat from one of the containers. Neither will two mixed types of gas spontaneously separate so that one of them can be found only in volume V_1 and the other only in volume V_2. As was shown in the foregoing Section, the processes that are observable in nature are accompanied by an increase in entropy. This leads us to the Second Law of thermodynamics:

> "Entropy can never be destroyed; it can only be created." (FALK and RUPPEL 1976:353, our translation)

If entropy could be destroyed, then processes like the ones described in the previous paragraph, e.g. the spontaneous separation of gases, would be possible in principle. The Second Law of thermodynamics excludes such processes. From the formulation of the Second Law we may conclude that temporally reversible processes are those that are not accompanied by a creation of entropy. In the course of such processes entropy remains constant. They can thus be reversed in time. On the other hand irreversible processes are those that are accompanied by a creation of entropy. The Second Law here excludes reversal with respect to time, for this reversal would be accompanied by the destruction of entropy. This, however, is excluded by the Second Law.

It is of course possible, however, to lower the entropy of a subsystem. Yet this can only happen in contact with the environment, which is where the difference in entropy will flow. Thus with the help of a pump the initial state, in which the gas is exclusively concentrated in one container while the other is evacuated, can surely be reinstituted. It is important to note that only the help of a pump, i.e. contact with the environment, makes this possible. Were the system to be enlarged in such a way that it contained the pump, its fuel, etc., then one would notice that the entropy of the total system has not diminished at all; it may have remained constant; most probably, however, it would have increased. It is thus possible to distinguish two parts of the change of entropy that accompanies a process. The first part dS_a depends on a contact with the environment, from which entropy is either absorbed or into which it is discharged. As a consequence this part can be either positive or negative. If the system chosen is sufficiently large, this part tends towards zero; an appropriately chosen system has no contact with its environment. Figuratively speaking, the system simply encompasses the environment. The other part dS_e describes the entropy created as a result of the irreversibility of the process under consideration. It is always greater than zero (cf. Falk and Ruppel 1976:247–8).

The formulation of the Second Law cited above is very suitable for the practical determination of whether a process can evolve in a particular direction or not. Its disadvantage is that it is not very descriptive, since it involves the notion of entropy. Other formulations that can be derived from the one cited by us have therefore been suggested. One example is Thomson's formulation of the Second Law:

> "It is impossible to construct a periodically working machine which does nothing but perform work and cool a heat reservoir." (FALK and RUPPEL 1976:367, our translation)

This means that it is impossible to construct a perpetuum mobile (a perpetual motion machine) (ibid.:368). In order to prove this proposition, let us consider a machine which transforms energy in the form of heat into energy in the form of work. Let us suppose it to be a system which is in contact with its environment. If this machine extracts energy dU_1 from a heat reservoir, it receives entropy to the amount of

$$dS = dU_1/T_1$$

– according to GIBBS' Fundamental Equation. T_1 here indicates the temperature of the heat reservoir. Since the machine works periodically, i.e. it is supposed to return to its initial state at some point, it must return the entropy received to the environment; for according to our formulation of the Second Law the machine is not able to destroy the entropy it received. This transmission of entropy, according to GIBBS' Fundamental Equation, always accompanies the transmission of energy to the amount of

$$dU_2 = T_2\,dS,$$

where T_2 denotes the temperature at which the transmission of entropy to the environment is to take place. This transmission of energy dU_2 to the environment, however, is nothing but the heating of a second system. This second system should be as cold as possible, i.e. it should have a small T_2, in order that the transport of energy into this system, $T_2\,dS$, be kept as low as possible. We thus see that the machine is only capable of transforming a certain part of the energy received, $T_1\,dS$, into work; the remainder $T_2\,dS$ must be released into a colder system.

Of practical interest is the relationship of the quantity of heat received to the work actually made available by the machine. In an electricity power plant, for example, this is the quantity of fuel needed for the electrical energy produced. This relationship can be easily calculated with the help of our above considerations. According to the First Law of thermodynamics mentioned in Sect. 3.2, energy is a conserved quantity. For a machine we thus obtain the result that the maximum amount of work extracted is equal to the difference between the quantities of heat received and released:

$$dA = dU_1 - dU_2 = (T_1 - T_2)\, dS$$

or

$$\eta = \frac{dA}{dU} = \frac{T_1 - T_2}{T} < 1.$$

This relationship is called the thermodynamic efficiency of a machine. It is always smaller than unity. A certain portion of the heat energy received is always released by the machine in the form of heat energy, at a lower temperature. Let us consider a power plant as an example. The heat is received from a vessel whose temperature is assumed to be 250 °C. Let the cooling water used for the release of heat have a temperature of 20 °C. We thus obtain for the efficiency

$$\eta = \frac{(250 + 273) - (20 + 273)}{250 + 273} = 0{,}44$$

Therefore this power plant can at best transform 44% of the energy input into electrical energy. The remainder must be released as waste heat.

The process described above is the process with the highest possible degree of efficiency, for it was assumed that no irreversible processes take place in the machine. Such irreversible processes within the machine would be accompanied by the creation of entropy. For a periodically working machine it would be necessary, however, to release this additionally created entropy into the environment thus increasing the temperature of the environment. This would increase the quantity of heat, however, which has to be fed into the cooling process. The efficiency would thus deteriorate.

3.5 The Irreversibility of Economic Processes and the Impossibility of the Land of Cockaigne

GEORGESCU-ROEGEN (1971:276) pointed out the close relationship between thermodynamics and economics. What SADI CARNOT (1824) was concerned with in his analysis of the ideal heat generating machine (CARNOT's circular process, cf. FALK and RUPPEL 1976:222–8) was really an economic problem; for he investigated the question: Under what conditions can a heat machine have a maximum output with a given heat input?

> "Extravagant though this thesis may seem *prima-facie*, thermodynamics is largely a physics of economic value ..." (GEORGESCU-ROEGEN 1971:276, his emphasis)

However much the neoclassical theory of economics is influenced by the procedure of physics, it is not influenced by thermodynamics but by classical mechanics (cf. e.g. GEORGESCU-ROEGEN 1971:2f., 40f., 334; MIROWSKI 1984). It is thus also characterized by the underlying reversibility of all processes. Physical considerations alone, however, suffice to show that the economic process, too, is a directional process; i.e. it cannot be circular (GEORGESCU-ROEGEN 1971:281). At the beginning of the fifties KOOPMANS (1951:48–50) explicitly considered the irreversibility of economic production processes in his path-breaking paper on the foundations of activity analysis. Since he utilized a linear technology – as do we in Chaps. 2 and 5 through 7 –, we shall briefly present his formulation at this point.
For this purpose we use KOOPMANS' (1951:36f.) terminology. Let the j-th production process (also called "activity") R_j ($j = 1, \ldots, J$) be a vector with coefficients a_{ij} ($i = 1, \ldots, I$), which indicate how many units of commodity i are needed or produced, if R_j is run with an intensity of $x_j = 1$. Contrary to our procedure (cf. Chap. 2) KOOPMANS uses negative coefficients for the factors of production. If all activities R_j are summarized in a $(I \times J)$-matrix A and the J-dimensional vector of process intensities is denoted by x, we obtain for the net output y the following relationship

$$y = Ax, \quad x \geqq O.$$

If a component of y is positive, then the corresponding commodity is produced in the economy characterized by A; if it is negative, then the corresponding good is used as an input. Irreversibility here means that it is impossible to run some or all processes at a positive level so that the total result of these processes, the net output vector y, is equal to the vector zero. For, were the vector zero producible, then the product of a production process could be used as an input in another process in order to produce the inputs originally utilized. In this case one could realize a production cycle without new inputs having to be added for its upkeep. This case is, however, excluded by the Second Law of thermodynamics. Thus there is no vector x with at least one positive component which fulfills the condition

$$y = Ax = O, \quad x \geqq o$$

(KOOPMANS 1951:48–9).

The irreversibility postulate can also be interpreted as an implicit acknowledgement of the temporal structure of production. For the production of a commodity cannot be reversed.

In the previous Section it was noted that one of the consequences of the Second Law of thermodynamics is the impossibility of a perpetuum mobile:

> "So, the popular economic maxim 'you cannot get something for nothing' should be replaced by 'you cannot get anything but at a far greater cost in low entropy'" (GEORGESCU-ROEGEN 1971:279)

KOOPMANS sees the impossibility of the Land of Cockaigne as the economic condition that reflects this fact: It is not possible to produce a good without an input. In the model presented above this statement is equivalent to the following one: There is no semi-positive vector x which fulfills

$$y = Ax \geqq O, \quad x \geqq O$$

(KOOPMANS 1951:50f.), i.e. for which at least one component of y is positive without any other being negative.

We have represented the postulates of irreversibility and of the impossibility of production without inputs in a linear approach. The formal relationship between the two conditions is shown by WITTMANN (1968:6). Formulations in models with non-linear technology can be found in textbooks of the theory of general equilibrium (cf. e.g. DEBREU 1959:39–42) and of production theory (cf. e.g. WITTMANN 1968:5f). It is our impression, however, that while irreversibility has begun to be accepted as an axiom, the physical meaning of such irreversibility has yet to be internalized into the conceptualization of many economists (FABER and PROOPS 1985, 1986). This lacuna has been recognized by SAMUELSON (1983:36):

> "In theoretical economics there is no 'irreversibility' concept, which is one reason why Georgescu-Roegen (1971) is critical of conventional economics."

3.6 Negative Flow of Entropy

As was shown in Sect. 3.4, it is one result of the Second Law that the entropy of a system can only be reduced when it is in contact with a second system (its environment). This second system assimilates the entropy that is discharged. What is important here is that this contact with the environment can be chosen in such a way that in the first system all important quantities such as the energy, the number of particles, and the volume remain unchanged. We shall now demonstrate this relationship in three examples.

Let us consider the earth as the system, the sun and the universe as the environment. The earth receives a constant flow of radiant (electro-magnetic) energy dU_S from the sun. Due to

$$dU_S = T_S \, dS_S$$

where T_S denotes the average temperature of the areas on earth that receives radiation from the sun. This flow of energy is accompanied by a flow of entropy dS_S to the earth. There is also a steady flow of heat and entropy

$$dU_E = T_E \, dS_E$$

from the earth into the rest of universe. T_E here represents the temperature of the areas that radiate. Since the temperature on earth is practically constant in the long run, this outflow of energy into the universe must compensate the inflow of energy from the sun

$$dU_S = -dU_E.$$

Thus we obtain for the difference of entropy on earth:

$$dS_S - dS_E = \left(\frac{1}{T_S} - \frac{1}{T_E}\right) dU_E < 0.$$

This difference of entropy, which results from the contact with the environment, is negative because the temperature T_S is greater than the temperature T_E. This difference in temperature corresponds essentially to that between day and night. The contact between sun, earth, and cosmos affects the balance precisely by a negative flow of entropy to the earth; all other thermodynamic quantities, however, remain unchanged.

As a second example, let us consider an economic system in contact with the environmental sector (cf. Fig. 1.1). Two volumes V_1 and V_2 of two different raw materials are directed into the economic system. There they are transformed into consumption goods, for example, and after a certain time returned to the environmental sector in the form of waste materials and pollutants. When they are returned, the two raw materials will generally no longer have their pure form; much more likely they will now be more or less mixed. Assuming that a complete mixing of the two substances takes place in the economic system, the process is described by the diffusion function which was explained in Sect. 3.3. According to Eq. 3.10, the outflow of entropy which accompanies a discharge of the material into the environment exceeds the inflow of entropy which accompanies the inflow of raw material into the economic system by

$$\Delta S = R\left(N_1 \ln(V/V_1) + N_2 \ln(V/V_2)\right).$$

Again it is important that the outflow of entropy is greater that its inflow. The contact with the environment thus makes it possible that the system discharges entropy created into the environment without any other quantity in the whole balance of the process being changed. This is a consequence of the fact that entropy is an extensive quantity; total entropy can thus be derived by the simple addition of the entropy of the economic system to that of the two raw materials. In this example the total quantity of raw materials in the economic sector is thus the same before and after the process. The contact between the two systems has thus had no other effect than a positive flow of entropy from the economic system. It is this outflow of entropy that makes economic activity possible. Or, put differently, a negative flow of entropy into the economic system was received from the environmental sector in the form of raw materials.

A third example is the inflow of fuels from the environment into the economic system. Let us assume that a particular volume V of a fuel (say coal) is channeled into the economic system where it is burnt with a temperature of T_V. According to GIBBS' Fundamental Eq. (3.7) – due to the relationship

$$dU_V = T_V\, dS_V$$

– a flow of entropy accompanies the resulting flow of heat; the changes of volume are negligible. As was explained in Sect. 3.4 by the example of the power plant, a part of this heat will be returned directly into the environmental sector. The part of energy that was transformed, for instance into electrical energy, will also be discharged into the environmental sector sooner or later, although the discharge of this energy will generally take place at a lower temperature T*. In the example of the power plant this is precisely the temperature of the cooling water. For the difference in entropy we thus obtain – just as we did in the first example –

$$\Delta S = \left(\frac{1}{T_V} - \frac{1}{T^*}\right) dU_V.$$

Once again all other thermodynamic quantities, such as the energy of the system or the number of particles, have remained constant. The residue from the burning of the fuel has been returned to the environmental sector in the form of emissions. On balance, the use of the fuel results in nothing but an outflow of entropy from the economic system. Or, put differently, the inflow of the fuel results in nothing but an inflow of negative entropy into the economic system.[15]

For simplicity we shall in the following make use of the notion of a negative flow of entropy. We wish to emphasize, however, that this notion is only appropriate when total systems are being considered in which all other thermodynamic quantities remain constant. The careful use of this notion allows us to avoid the more precise, but also more difficult, formulation in terms of balance equations.

3.7 Entropy, Order, and Information

We have introduced the notion of entropy as an extensive correlate to the intensive quantity temperature. This definition corresponds to the procedure explained in Sect. 3.1, which was to find physical quantities which are appropriate for the description of thermodynamic systems. The definition of the notion of entropy thus became necessary for formal reasons.

With the help of physical quantities we describe properties of systems which are known by experience. Thus everybody has some idea what properties are described by the physical quantities force, temperature, or speed. As a result of the axiomatic, formal definition of entropy, however, we cannot correlate with it any intuitive properties. So far, the quantity of entropy is thus nothing but a purely formal quantity for us. However useful this quantity may be in the context of the Second Law, it simply has no descriptive meaning. Systems that are free of interdependencies, like ideal gases and dilute solutions, at least present the possibility of correlating a property of the system with the notion of entropy. This is what BOLTZMANN showed when he noticed that entropy can describe the order of a system. For a better understanding of this, let us reconsider the diffusion of gases. As was described in Sect. 3.3, the diffusion of gases is accompanied

[15] One could make the case that the main function of the fuel substance is not to supply energy, but to withdraw entropy from the economic system. The heat engine, which is the method usually chosen in practice, represents nothing but a particularly effective way of creating this negative flow of entropy.

by an increase in entropy. The initial state of this process, in which the two gases are separate from each other in their respective cylinders, is more orderly than the final state, in which the gases are perfectly mixed (cf. Fig. 3.2). The increase in entropy is thus paralleled by a decrease of the level of order of the system. In systems that are free of interdependencies, entropy describes the system's level of order – with low entropy signifying a high level of order and high entropy a low level of order. Entropy thus describes the degree of disorder of a dynamic system free of interdependencies. As the example of a spontaneously separating water-oil-mixture (cf. Sect. 4.1.2 below) shows, this interpretation of the notion of entropy is strictly limited to systems that are free of interdependencies.

Another approach to the notion of entropy that is totally different from the introduction here chosen relies on information theory. In this approach, which goes back to SHANNON, entropy is a measure for the information contained in a message. Since in the following we shall not refer to the resulting interpretation of entropy, we shall not here take a closer look at the information-theoretic interpretation of the notion of entropy.[16] Suffice is to note that the thermodynamic interpretation of entropy can be derived from the information-theoretic interpretation of entropy (cf. STUMPF and RIECKERS 1976, PROOPS 1985, 1987).

3.8 Using the Entropy Approach to Characterize the Environment as a Recipient of Pollutants

The function of the environment as a recipient of pollutants on the one hand and as a supplier of resources on the other are the two areas to which we shall apply the entropy approach in this Section and in the next Chapters. In characterizing the environment as a recipient of pollutants, we shall restrict ourselves in this Section mainly to

a) an explanation of the notion of "ecological equilibrium" with the help of the entropy approach, and

[16] See GEORGESCU-ROEGEN (1971, Appendix B:388–406) who discusses the relation between ignorance, information, and entropy.

b) to a description of harmful effects of pollutants on the environment in terms of changes of the entropy in the environmental sector.

In the next Chapter we shall show more extensively that the extraction of resources leads to changes of entropy in the environmental sector also. Thus both functions of the environment will receive a common theoretic foundation on the basis of the entropy approach.

It is often postulated that it should be one of the most important aims of environmental protection measures to maintain the ecological equilibrium of nature.[17] Mostly, however, an appropriate measure with which to appraise and estimate this complex quantity is missing. In what follows we shall show that the notion of entropy can be used to describe the environment as a recipient of pollutants (Sect. 3.8.1) and to identify deviations from ecological equilibrium[18] as an change in entropy.[19]

3.8.1 Entropy and Equilibrium

In explaining Fig. 1.1 in Chap. 1 it was already mentioned that our ecological system is an open system. In Sects. 3.4 and 3.6 we showed further that the openness of a system is an essential prerequisite for an evolutionary development to run contrary to the general development of the total system. For since the total system itself converges towards the thermodynamic equilibrium in which all differences in temperature have been eliminated[20] the state of equilibrium of an open system can only be an inter-

---

[17] Cf. the Environmental Program of the German government (1971:35).

[18] In ecology, the notion of an ecological equilibrium is mostly used in the following sense. It is the state that corresponds for instance to the undisturbed ecological status quo of a biotope. As a consequence it contains a normative aspect, just as the equilibrium existing in an economy often used to be regarded as a state to be preserved in economic textbooks. – Whenever we employ the notion of an ecological equilibrium in what follows we shall do so without its normative aspect.

[19] We conjecture that it may be possible to use the notion of entropy as a direct measure for an ecological equilibrium. Lengthy and fundamental considerations would have to be entered into, however, in order to verify this conjecture.

[20] Overall thermodynamic equilibrium is thus also refered to as the "heat death" of the universal system.

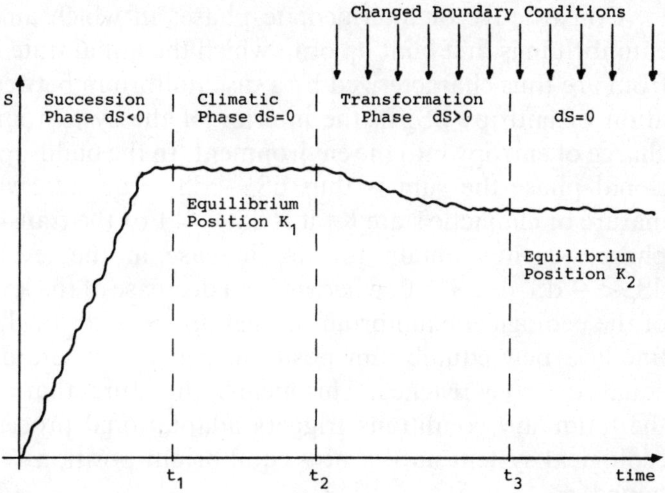

Fig. 3.3. Schematic presentation of the equilibrium of a developing ecosystem for changing boundary conditions

mediate state. Yet for a certain amount of time this intermediate state can be seen as a state of equilibrium. It is characterized by KREEB (1979:93) as follows:

> "It is thus more accurate to speak of a *flow equilibrium* or 'steady state' (stationary state). It specifies the preservation of a system in a certain condition (informational level, order) in spite of the constant making and unmaking of the partial structures (individuals), and it actually corresponds to a *disequilibrium state of the system*." (his emphases, our translation)

Since entropy is a measure of the level of order of a system there exists the possibility to specify the ecological equilibrium of a system as a state in which the creation of entropy, dS_e, and the negative flow of entropy, dS_a, are balanced, i.e. $dS = dS_e + dS_a = 0$.[21] Let us consider the settling in an ecological habitat (cf. Fig. 3.3)[22] as an explanatory example.

[21] To avoid misunderstandings concerning the notion of an ecological equilibrium we refer the reader to footnote 18. Note, however, that even if the entropy of the system as a whole is constant, there might be subsystems with de- or increasing entropy.

[22] In constructing this figure we followed the presentation of the German Council of Experts for Environmental Questions (1978:23–5) which contains a survey of possible disturbances and adaptive processes in the ecosystem.

Constructive or intermediate phases in which an ecological equilibrium is first built up or in which the initial state is deviated from are thus characterized by a disequilibrium between the creation of entropy dS_e in the interior of the system and the discharge of entropy into the environment. In the build-up or successional phase the sum is thus $dS_e + dS_a = dS < 0$, while in the mature or climactic state K_1 it is $dS = 0$. For the transformation-phase we thus obtain for an increase in the level of order $dS_e < -dS_a$ or $dS < 0$; whereas for a decrease of the level of order of the ecological equilibrium we get $dS_e > -dS_a$ or $dS > 0$ until finally a new equilibrium position, e.g. K_2, where dS is again equal to zero is reached. This means, therefore, that a change of the boundary conditions triggers adaptational processes in the ecological system until a new equilibrium position is finally attained.

3.8.2 The Entropy Approach as a Means of Determining Deviations from Equilibrium

Effects on the equilibrium (or on the level of entropy) of a certain habitat within the environmental sector can, for instance, stem from emissions. In the environmental models of the first and second Chapters the relationship between emissions, pollution, and the flow of environmental goods was represented with diffusion and damage functions.[23] In our environmental models these functions show how emissions are distributed through diffusion and transformed by natural processes. The remainder, i.e. the pollutants, negatively affects the flow of environmental goods according to the damage function.

With the entropy approach these relationships between emissions, pollution, and the flow of environmental goods (cf. arrows 6, 7, and 8 in Fig. 1.1) can be described as follows. In the course of production and consumption an economic system wears down capital goods and simultaneously produces emissions. Production and consumption thus change the level of order of an economic system so that the entropy of the system is increased. Were

[23] Cf. Eqs. (1.9), and (1.18) in the first Chapter, for instance, or the explanations in Sect. 2.1.2.

this permanent creation of entropy in open economic systems not counteracted by some other flow of entropy, a state of equilibrium could not be maintained in the long run. Such a negative flow of entropy is secured by the flow of environmental goods into the economic system. An example of such a negative flow of entropy is the energy of the sun that is flowing to the earth (cf. Sect. 3.6) and the related ability of nature to degrade emissions in the environmental sector.

To bring this out more clearly, let us assume an initial ecological equilibrium. Now let the production of consumption goods be started with wastes being emitted. If these emissions, which impair the flow of environmental goods, are not counteracted, they will create additional entropy in the environmental sector which is not compensated by a negative flow of entropy from the outside. Thus the state of equilibrium cannot be maintained.

3.9 A First Attempt to Employ Entropy as a Variable in a Pollution Function[24]

Lack of energy may not only stop economic growth, but may even lead to a decline of economic activity. This may happen in developing countries, which carry through a deforestation of 20 million hectares each year (GLOBAL 2000, 1981:126), an area as half as large as the State of California[25]. But let us assume that other energy sources are innovated fast enough in such a way that sufficient energy can be supplied. Then we still have to cope with the problem of environmental pollution.

Consider an economic system which uses two different resources of volumes V_1 and V_2 respectively. They are used to produce consumption goods, which in the course of time wear out or are consumed. In any case, they will pollute the biosphere in the form of waste and will finally increase entropy. For the sake of illustra-

[24] This Section is based on Sect. 3.1 of FABER (1985) who in turn draws on KÜMMEL (1980).

[25] We note that developed countries also deforested when they were in their stage of agricultural and industrial development. This statement holds in particular for Great Britain.

tive purposes let us assume that the two resources are completely mixed in production and that they behave like the two ideal gases in the example mentioned in Sect. 3.3. We then obtain according to (3.10)

$$\Delta S = R(N_1 \ln(V/V_1) + N_2 \ln(V/V_2)) > 0 \tag{3.11}$$

which shows the increase of entropy in the environment. Hence there is a continuous flow of entropy into the environment, which is generated by the economic production process. We note that matter and the energy are conserved during the manufacturing process, according to the First Law of thermodynamics, but that the entropy increases according to the Second Law.

We may use entropy as a highly aggregated measure of pollution. Let V be the corresponding ecological space, then a total measure can be defined as

$$W = \frac{1}{V}\frac{dS}{dt} \tag{3.12}$$

where dS/dt is the increase of entropy in time (for the following derivation of (3.13) and (3.14) see KÜMMEL 1980:25–29). We can define such a measure W_i for every pollutant i. To construct a pollution function we need critical values W_{ci} of the degrees of pollution, such that society cannot tolerate $W_i > W_{ci}$. Of course, different types of pollutants have different values. Thus the ones for heavy metals like mercury or cadmium, which are very dangerous for any kind of life, will be comparatively very low.

Since the earth is not a closed system we must allow for the possibility of natural purification. Denoting the corresponding purification rates by W_{oi}, we can formulate a pollution function for each type of pollutant

$$p_i = p_i(W_i; W_{ci}, W_{oi}), \tag{3.13}$$

which measures how much pollutant i contributes to the loss of welfare for society.[26] This indicator should have the following properties:

[26] This interpretation is in value terms while KÜMMEL's is in real terms and concerns the impact of the pollutants on growth.

1. Its value is zero or low if $W_i \ll W_{ci}$ and it is infinity if $W_i > W_{ci}$;
2. the larger is W_{oi} the less the approach of W_i to W_{ci} will increase p_i.

The sum over all pollutants i = 1, ..., M

$$p = \sum_{i=1}^{M} p_i(W_i; W_{ci}, W_{oi}) \tag{3.14}$$

can be used as a total pollution function. We note that if the value of just one type of pollution is equal to its critical value the total pollution function will be at a maximum.

Let us turn now to a particular pollutant, viz. thermal pollution. For heat the critical value W_{ci} is $3 \cdot 10^{14}$ Watts (KÜMMEL 1980, p.110). If the annual growth of the industrial world production is 5% this limit will be reached (under certain assumptions) around 2050 and if it is 3% around 3000 (KÜMMEL, 1980, p. 111).

Heat is a ubiquitous pollutant; it results as a by-product not only of the manufacturing processes of capital and consumption goods, but also of environmental protection activities. Thus, for example, the water treatment plant of the BASF Ludwigshafen uses as much energy as a city with 50,000 inhabitants. The corresponding thermal pollution is not dealt with. As KÜMMEL (1980, p. 110) pointed out, we do not yet have a feasible technology to get rid of a thermal pollution of the magnitude of the heat barrier of $3 \cdot 10^{14}$ Watts mentioned above. These considerations imply that thermal pollution may finally be the most severe form of pollution.

4 Using the Entropy Approach to Characterize the Environment as a Supplier of Resources

So far the flow of environmental goods (or the negative flow of entropy) from the ecological system into the economic system has not been closely specified. In the following we want to show how the problem of resource scarcity can be specified within the entropy approach. We start by taking note of the fact that the flow of environmental goods, carrying with it a negative flow of entropy into the economic system, contains also the raw materials needed for the manufacturing of the consumption and capital goods in the production sector. The supply of the economic system with raw materials from the environmental sector can therefore only be secured if the ecological system possesses the long-run capability of providing a negative flow of entropy into the economic system. This negative flow of entropy must correspond to the desires of the economic agents.

In Sect. 4.1 we shall first derive a relationship between factor inputs in the extraction of resources, and resource concentration. We shall show more specifically that the separation process in the extraction of resources can be described as the reversal of a process that is irreversible in nature – the process of diffusion. The reversal of this diffusion process is caused by the employment of a flow of negative entropy. In the following we shall formally derive the relationships between the change of entropy, the input of factors in resource extraction, and the concentration of resources.

In Sect. 4.2 we shall use these relationships in order to show that any extraction of resources in the environmental sector leads to the creation of entropy. This can be reduced to some extent, however, by preventive measures such as recycling or the controlled disposal of waste materials.

In Sect. 4.3 we shall comment on the limits of our approach.

4.1 Resource Concentration and Factor Requirements in the Extraction of Resources

4.1.1 The Separation Process as a Reversal of the Diffusion Process

The process of resource extraction can be divided into two phases: the appropriation process and the succeeding separation process. During the appropriation process (cf. arrow 1 in Fig. 4.1) volume elements V which contain a certain type of elements m – e.g. iron – needed in production are brought from the environmental sector into the economic subsystem. During the following separation process, the second phase, the desired raw material is separated from the other substances within the base material, as in the case of the extraction of iron from iron ore. In this concentrated form the raw material m is processed further (e.g. raw iron into steel) and is finally used in the production of consumption or capital goods (arrow 2). Those substances that are not needed are channeled back into the environmental sector (arrow 3).

The purpose of separation is the purification of raw materials, which is required for their use as a factor of production. This means that a raw material m required for production has to be

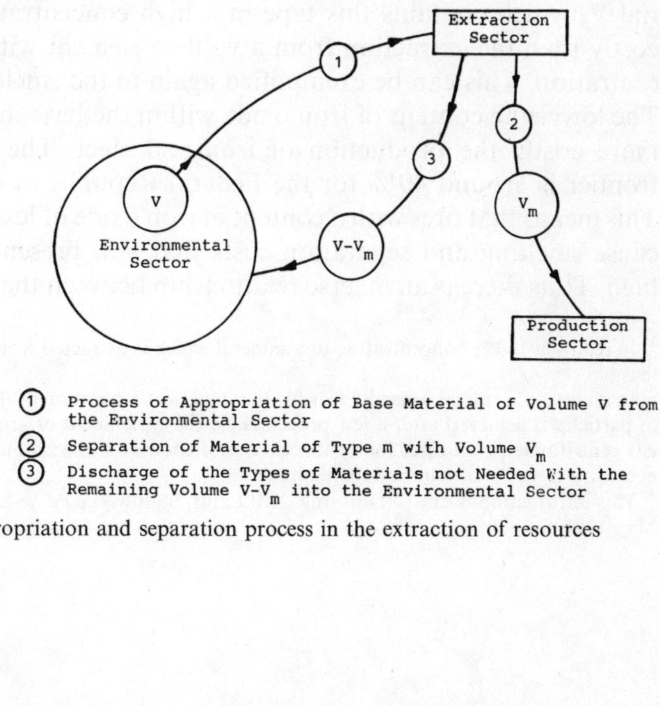

Fig. 4.1. Appropriation and separation process in the extraction of resources

separated as thoroughly as possible from those other substances which are simultaneously put at the immediate disposal of the economic system by the appropriation process. For the formal description of extraction let us first define the resource concentration of elements of type m. Let $N_1, \ldots, N_M$ be the quantities of mols of the resources of types $1, \ldots, M$ which are found in one unit base material V. We thus obtain for the resource concentration of the materials of type m in V

$$\kappa_m(V) = \frac{N_m}{\sum_{j=1}^{M} N_j} \qquad m \in \{1, \ldots, M\}. \tag{4.1}$$

When separating, one maximizes concentration under the restriction of economic considerations. Only for the borderline case of a 100% separation[1] is a resource concentration of $\kappa_m(V_m) = 1$ achieved within the volume portion V_m, which contains the elements of type m in concentrated form. This was assumed for the transition from state 1 to state 2 in Fig. 4.2. In this special case V_m thus contains nothing but elements of type m.[2] If separation is imperfect, however, the concentration of resources achieved in V_m is smaller than one, but larger than the initial concentration $\kappa_m(V)$. Empirical observation suggests the assumption that the separation of particles of type m from an appropriated base material V, which contains this type in a high concentration, is less costly than the extraction from a volume element with low concentration. This can be exemplified again in the smelting of ore. The lower the content of iron oxide within the base material, the more costly the production of iron and steel. The rentability frontier is around 40% for the Federal Republic of Germany.[3] This means that ores with a content of iron oxide of less than 40% cause smelting and separation costs that – at present – are too high. Thus there is an inverse relationship between the separation

---

[1] In reality a 100% concentration of a mineral which is extracted from an ore is not feasible.

[2] In reality, as a rule, it is hardly feasible that a perfect separation of the desired type of particles is achieved after a few processes. In the enrichment of uranium with the gas centrifugue procedure, for instance, several thousand extraction processes are necessary in order to win fissionable material.

[3] This information is due to Dipl.-Ing. H.P.Lauff, Stahlwerke Peine-Salzgitter A.G., May 1981.

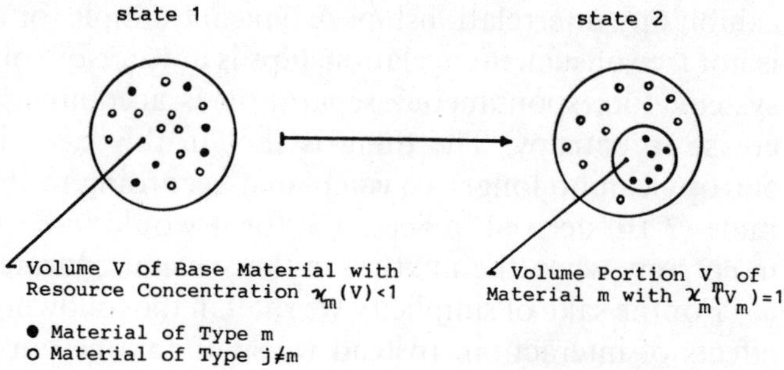

Fig. 4.2. The change of resource concentration during the separation process

cost for resources and the initial concentration in the appropriated material.

It would now be a simple procedure to introduce a relationship between resource concentration and factor input that is based on this empirical observation into a resource model by way of an assumption. We shall here not use such a procedure, however, because the ultimate causes for this relationship would remain obscure. Instead, we are going to use a relatively complex method, i.e. the entropy approach. It is our opinion that this effort pays dividends, because it provides a common basis for the different forms of uses of the environment. We also believe that our procedure helps in working out the real side of resource extraction, especially the relationship between resource concentration and the energy requirements for extraction.

4.1.2 The Change in Entropy as a Function of Resource Concentration

In the following we shall derive the change in entropy as a function of resource concentration. As we can see from Figs. 4.1 and 4.2, the separation of resources in question is a direct reversal of the diffusion process described in Sect. 3.3. As explained in Sect. 3.4, it is a consequence of the Second Law of thermodynamics that, in a closed system, separation can never take place by itself, for such a process is accompanied by a decrease in entropy.

Yet in this simple form the statement is true only for substances that do not interact with each other, i.e. that do not exhibit any interrelationship. A typical example for a system that is not free of such interrelationships is a water-oil mixture. In this system, too, spontaneous separation is accompanied by an increase in entropy. The thing is that in this case the change in entropy can no longer be calculated according to the simple formula (3.10) derived in Sect. 3.3, for it would indicate a decrease in entropy when the mixture in the system separates.

For the sake of simplicity we shall in the following ignore such effects of interaction. Instead we shall confine ourselves to systems in which the interrelationships can be ignored. This assumption represents a considerable limitation for our model in its present state; for in reality only gases and dilute solutions fulfil this assumption.

We would now like to supplement our approach, as developed in Sect. 3.3, in the following manner. We assume, that the extraction process does not change the total volume. We further assume that each resource possesses a certain volume whose relationship to a mol of the substance is the same for all substances. Our approach is thus essentially a mixture of a simple solid body model and a gas model.[4]

Let us now take a volume V which contains the resources $m = 1, \ldots, M$ with the number of mols N_m. Let the number of mols of the total system be

$$N = \sum_{m=1}^{M} N_m . \tag{4.2}$$

According to (4.1) we thus obtain for the concentration κ_m of the resource m

$$\kappa_m = \frac{N_m}{N} \qquad m \in \{1, \ldots, M\}. \tag{4.3}$$

[4] The first postulate describes that there is a linear relationship between the volume of a solid body and its mass (which is given by the specific density). The second postulate is known in gas theory as the Rule of AVOGADRO. According to this rule a given volume of gas always contains the same number of mols regardless of which gas is chosen and provided that conditions are the same.

Each resource has the volume V_m for which, according to our assumption, the following is valid:

$$V_m = \frac{N_m}{N} V = \kappa_m V$$

$$\frac{V_m}{V} = \frac{N_m}{N} = \kappa_m . \tag{4.4}$$

We further assume that the resources are evenly distributed within volume V, i.e. that we are dealing with a perfect mixture. Let us now consider the following case: from volume V we want to separate a resource fully, i.e. we want to condense this resource into volume V_1. Let this resource have the index 1. Since resource 1 fully occupies volume V_1 – V_1 is its own volume – the other resources m, m = 2, ..., M, must be contained in volume $V - V_1$, as the total volume is assumed constant in our model. The difference in entropy arising in separation is obtained analogously to the case of the diffusion of two gases. Only the sign is different, since we are dealing with the reverse process. In analogy to (3.9), we thus obtain for the change in entropy of resource 1 which is condensed from volume V into volume V_1

$$\Delta S_1 = - N_1 R \ln (V/V_1) \tag{4.5}$$

and for the other resources, which are condensed from volume V into volume $V - V_1$

$$\Delta S_m = - N_m R \ln (V/(V - V_1)) \quad (m = 2, \ldots, M). \tag{4.6}$$

Recalling that entropy is an extensive quantity, i.e. the difference in entropy of the total system is nothing but the sum of the individual differences, and that, according to (4.2)

$$\sum_{m=2}^{M} N_m = N - N_1 , \tag{4.7}$$

we obtain in analogy to (3.10)

$$\Delta S = \sum_{m=1}^{M} \Delta S_m \tag{4.8}$$

$$= - R\{N_1 \ln(V/V_1) + (N - N_1)\ln(V/V - V_1))\}^5$$

or

$$\Delta S = N_1 R\{\ln \kappa_1 + ((1 - \kappa_1)/\kappa_1)\ln(1 - \kappa_1)\}. \tag{4.9}$$

This entropy difference is still related to the volume V, or the total number of mols N. It is the entropy difference necessary to produce one unit of the desired resource, however, which is of economic interest. In our example, the quantity 'mol' suggests itself as a quantitative unit (remember that N was measured in mol). We thus define the mol-specific entropy difference

$$S_{N_1} = \Delta S/N_1 = R\{\ln \kappa_1 + ((1 - \kappa_1)/\kappa_1)\ln(1 - \kappa_1)\} \tag{4.10}$$

which indicates the negative flow of entropy necessary to extract one mol of resource 1. The resource concentration κ_1 lies within the interval $0 < \kappa_1 < 1$, which yields $0 < (1 - \kappa_1) < 1$. Since the logarithmic function takes on negative values for this interval, the entropy difference ΔS_{N_1} is also negative, as was to be expected from the analysis in Sects. 3.4 and 3.6.

The entropy difference of other extraction aims, such as the complete separation of all resources, can be calculated in a similar fashion.

The two Figs. 4.3 and 4.4 show how the mol-specific entropy difference S_{N_1} is related to the initial concentration κ_1. The Figures represent the numeric values (our calculations) reprinted in the Appendix to this Chapter. Fig. 4.4 shows the behavior of the mol-specific entropy difference as a function of the initial concentration; Fig. 4.3 shows its marginal change, i.e. the first derivative with regard to the initial concentration. As we can see from Fig. 4.4 , the negative flow of entropy accompanying the

[5] A similar formula is derived by Chapman and Roberts (1983:91) for metals. The two authors also discuss the difficulties of applying this formula in practice.

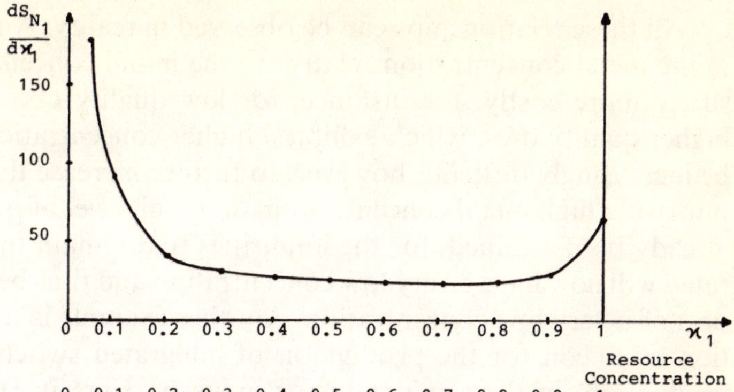

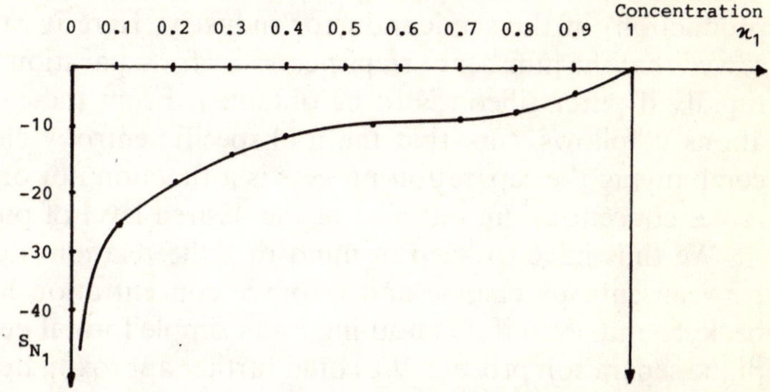

Fig. 4.3 (*above*). First derivative of the mol-specific entropy change with respect to resource concentration

Fig. 4.4 (*below*). Mol-specific entropy change as a function of resource concentration

extraction process is smaller the higher is the initial concentration.[6]

As is indicated in Fig. 4.3, the first derivative increases beyond all positive bounds both for very low and for very high initial concentrations. We can also say that, in marginal terms, the negative entropy flow accompanying the extraction process necessary to increase the resource concentration by one marginal unit must be higher for very low or very high initial concentrations than for intermediate ones.

---

[6] This confirms the first derivation of equation (4.10) with regard to the initial concentration. For

$$d(\Delta S_{N_1})/d\kappa_1 = R\left\{\frac{1}{\kappa_1} + \frac{-\ln(1-\kappa_1)-\kappa_1}{\kappa_1^2}\right\} = \frac{-R}{\kappa_1^2}\ln(1-\kappa_1).$$

All these relationships can be observed in reality. An increase in the metal concentration relative to the initial concentration is surely more costly, for instance, for low quality ores than for higher quality ores, which exhibit a higher concentration. It will be increasingly difficult, however, to further increase the level of purity if a high metal concentration, or a high level of purity, has already been attained, for the impurities that remain to be separated will now have a very low concentration and thus behave like ores of a very low concentration. Another example is the extraction of silicon for the production of integrated switches (chip-production) in the semiconductor industry. Here it can be observed that the number of steps necessary for separation increases rapidly if purer silicon is to be obtained. From these considerations it follows, too, that the mol-specific entropy change accompanying the separation process is a function not only of the initial concentration, but also of the desired level of purity.

We thus have to keep in mind that the relationship derived between entropy change and resource concentration has a real background, even if it is nothing but a simple formal description of the separation process. We could further approximate reality if we also took into consideration the binding forces between the types of particles to be separated.

4.1.3 The Energy Requirements in the Extraction Process

In Sect. 4.1.1 we explained that the separation process can be interpreted as the reversal of a diffusion process. Separation processes, however, lead to a reduction in entropy. According to the Second Law of thermodynamics this is only possible if there is a negative flow of entropy into the system.

Let us denote this flow of entropy by ΔS_a. It differes from the change of entropy aimed at in the separation process, ΔS_e, only by the sign. The change of entropy required for the separation process can be obtained from (4.8); we thus get

$$\Delta S_a = -\Delta S_e = -R\{N_1 \ln \kappa_1 + (N - N_1)\ln(1 - \kappa_1)\}. \tag{4.11}$$

If we relate the flow of entropy ΔS_a to the mass unit already utilized in (4.10), we obtain for the mol-specific entropy flow in the diffusion process

$$-\Delta S_{N_1} = -R\left\{\ln \kappa_1 + \frac{1-\kappa_1}{\kappa_1}\ln(1-\kappa_1)\right\}. \qquad (4.12)$$

This mol-specific entropy flow must be fed from the environment into the open system, if resource 1 is to be separated therein.

We would now like to demonstrate the relationship that exists between this flow of entropy and the accompanying exchange of energy between the system and its environment. We here utilize GIBBS' Fundamental Equation which was introduced in equation (3.7) above. Let us assume that every form of energy dU fed into the system is used exclusively for an increase of the level of order of the system. GIBBS' Fundamental Equation can thus be simplified as follows:[7]

$$dU = T dS. \qquad (4.13)$$

Integration of (4.13) yields the simple difference equation

$$\Delta U = T\Delta S. \qquad (4.14)$$

It shows how the flow of entropy is linked to the accompanying flow of energy.

If the extraction process fulfills the same assumptions as were made in the derivation of (4.14), then, taking into consideration (4.14) and (4.11), (4.12) yields the following relationship for the mol-specific energy requirements U_{N_1}

$$\Delta U_{N_1} = \frac{1}{N_1}\Delta U = \frac{T}{N_1}\Delta S_a = -\frac{T}{N_1}\Delta S_e \qquad (4.15)$$

$$= -RT\left\{\ln \kappa_1 + \frac{1-\kappa_1}{\kappa_1}\ln(1-\kappa_1)\right\}.$$

[7] We remind the reader that according to our comment on formula (3.8), the correct relationship should be $T dS = dU - |\partial A|$ instead of (4.13). ∂A here is the work which has been performed on the corresponding system. Since we have assumed, however, that the total energy is used to change the order of the system only, $|\partial A| = 0$. We employ this simplified relationship since we are mainly interested in qualitative results.

Comparing the energy change U_{N_1} in (4.15) with the change in entropy S_{N_1} in (4.10) which is caused by it we realize that both expressions differ in the temperature T only. Assuming T to be constant both expressions are similarly dependent on resource concentration κ_1. The mol-specific energy requirement U_{N_1} increases with a lowering of the initial concentration κ_1. The marginal increase of the mol-specific energy requirement is larger both for low and high resource concentration than it is for medium resource concentration.

In order to demonstrate the functional dependence on resource concentration to the reader, we pick a quantitative example. Let the temperature T for simplicity have the value 0 °C (equal to 273° Kelvin). Any other choice of T would do nothing but change the scale of the ordinate in Fig. 4.5. The factor preceding the brackets in (4.15) has the value

$$RT = 2269.8312 \frac{\text{Watt sec}}{\text{Mol}} \tag{4.16}$$

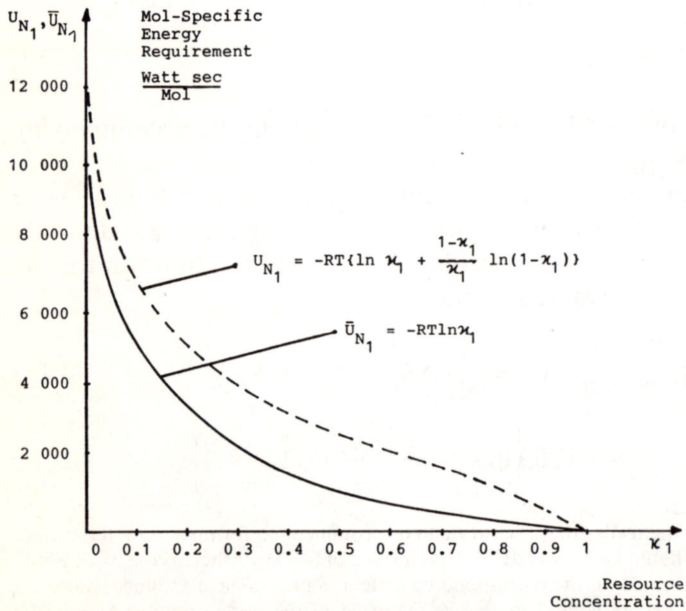

Fig. 4.5. Relationship between energy requirement and resource concentration in the extraction process at T = 273° Kelvin according to Eqs. (4.15) and (4.17)

with the dimension Watt · sec/mol valid for the mol-specific energy requirements.

In the following Section we shall investigate in which form the mol-specific energy requirement U_{N_1} is rendered. For simplicity we shall, however, use the lower bound

$$\overline{U}_{N_1} = - RT \ln \kappa_1 \tag{4.17}$$

for the mol-specific energy requirement. For (4.15), according to (4.17), has a lower bound, i.e.

$$U_{N_1} > \overline{U}_{N_1}. \tag{4.18}$$

Figure 4.5 shows that the mol-specific energy requirement increases more than proportionately with decreasing initial concentration of the desired type of material 1.

4.1.4 Factor Input in Resource Extraction

For simplicity we assume that the total mol-specific energy (i.e. its lower bound) required for the extraction of resources, consisting of appropriation and separation, is being rendered exclusively in the form of manual human labor. The lower bound of the mol-specific energy requirement in (4.17) then indicates the lower bound of the mol-specific labor input into the separation process. Denoting the mol-specific labor input into the separation process by $l_1^E(\kappa_1)$ we obtain

$$l_1^E(\kappa_1) = - \varepsilon_1 \ln \kappa_1 . \tag{4.19}$$

In order to simplify our notation we have in (4.17) abbreviated the coefficients by $\varepsilon_1 = RT$.

We now turn to the appropriation process (cf. Fig. 4.1). The mol-specific labor input into the appropriation process indicates what amount of labor is necessary in order to bring one mol of a mixture of substances from the environment into the economic system.

Let us consider a deposit of resource 1 with the concentration κ_1. If we appropriate from this deposit the amount of N mols, containing $N_1 = 1$ mol of resource 1, then

$$\kappa_1 = \frac{N_1}{N} = \frac{1}{N}.$$

$1/\kappa_1$ mols are needed in order to win one mol of the resource. If the labor input necessary to appropriate one mol of the mixture of substances from the deposit is denoted by α_1, then

$$l_1^A = \alpha_1/\kappa_1 \tag{4.20}$$

units of labor are necessary in order to appropriate $1/\kappa_1$ mols of the mixture of materials, from which one mol of the resource is to be finally extracted. For the mol-specific labor input in appropriation and extraction we thus get from (4.19) and (4.20) a total of

$$l_1^R(\kappa_1) = l_1^A(\kappa_1) + l_1^E(\kappa_1) = \alpha_1/\kappa_1 + (-\varepsilon_1 \ln \kappa_1). \tag{4.21}$$

Forming the first derivative of (4.21) with respect to the resource concentration we obtain

$$\frac{\partial l_1^R(\kappa_1)}{\partial \kappa_1} = \frac{-\alpha_1}{(\kappa_1)^2} + \frac{-\varepsilon_1}{\kappa_1}. \tag{4.22}$$

We recognize that, since $\kappa_1 < 1$, the first term of (4.21) increases considerably faster with falling resource concentration than does the second term (cf. Fig. 4.6). For the borderline case of $\kappa_1 = 1$ the separation process is unnecessary; for this reason $l_1^E = 0$ for $\kappa_1 = 1$.

From our representation of the process of extraction of resources on the basis of the entropy approach we recognize how total labor input $l_1^R(\kappa_1)$ depends on the concentration κ_1, the value of ε_1, and the deposit-specific value of α_1. These substance- and deposit-specific values differ from resource to resource.[8]

[8] An empirical illustration of some micro- and macroeconomic aspects of the relationship (4.19) and (4.21) is given by FABER and WAGENHALS (1987, Sect. 3), who analyze data of a single copper mine in Chile and the total copper mine production of the United States.

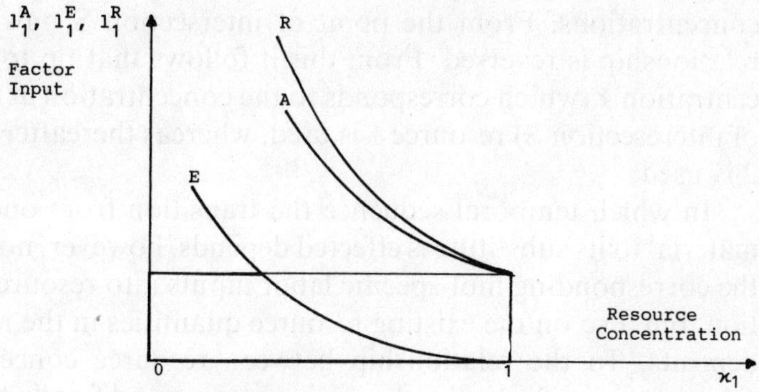

Fig. 4.6. Factor input in resource extraction

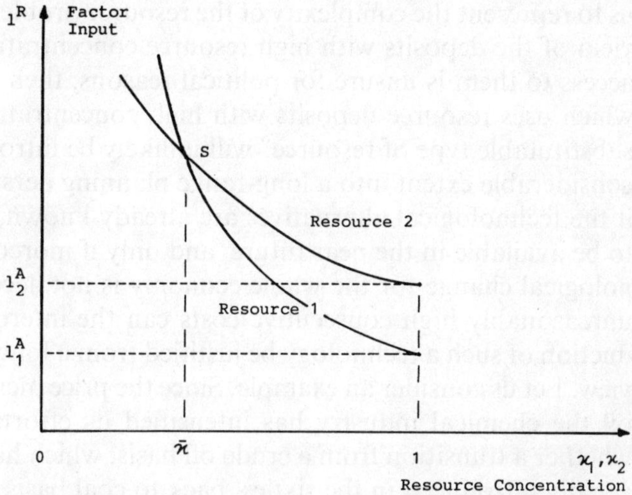

Fig. 4.7. Factor input in resource extraction for two different types of resources

This observation prompts us to investigate whether the entropy approach can help us to decide when resources are substituted. For this purpose we assume that there are two types of resources, 1 and 2, which are perfect substitutes in the production of a good. Each can be found in a different deposit. In Fig. 4.7 are represented different curves for this situation. If we now interpret both curves in Fig. 4.7 for decreasing concentrations we realize that, up to the point of intersection S, the mol-specific labor input for resource 2 is greater than for resource 1, for identical resource concentrations. From the point of intersection S onwards this relationship is reversed. From this it follows that up to the concentration $\bar{\kappa}$ (which corresponds to the concentration at the point of interesection S) resource 1 is used, whereas thereafter resource 2 is used.

In which temporal sequence the transition from one type of material to its substitute is effected depends, however, not only on the corresponding mol-specific labor inputs into resource extraction, but also on the existing resource quantities in the respective deposits. To the relationship between resource concentration, energy requirements, the change in entropy, and finally the factor input into resource extraction is thus added the quantity distribution of the stocks of resources. Only both aspects together allow us to represent the complexity of the resource problem. For, if the yield of the deposits with high resource concentration is low, or access to them is unsure for political reasons, then a technology which uses resource deposits with high concentration of a non-substitutable type of resource[9] will unlikely be introduced to any considerable extent into a long-range planning perspective. Only if the technological alternatives are already known, or are likely to be available in the near future, and only if moreover the technological change for the whole economy is not accompanied by unreasonably high consecutive costs can the intermittent introduction of such a technology be justified from a long-run point of view. Let us consider an example. Since the price increase of crude oil the chemical industry has intensified its efforts to examine whether a transition from a crude oil basis, which had been fairly rapidly introduced in the sixties, back to coal basis is technically possible, and whether it will be economical in the near future from a private profit perspective. It cannot be excluded that the conversion to a crude oil basis will turn out to be nothing but an "episode" for the chemical industry. In the long-run, and from a social perspective, it may even turn out to have been a suboptimal development during which too much crude oil was used up.

We can interpret the stocks of resources and their distribution among the respective deposit sites as an ecological "advance" or

[9] It may be tempting to argue that resources which cannot be substituted are essential, i.e. that $F(K, L, R = 0) = 0$. As Dasgupta and Heal (1979:196ff.) show, however, this would be a mistake.

pre-payment which is being tapped when economic activities are pursued.

The extraction of resources influences their distribution and thus the state of the environmental sector. Regardless of whether the environmental sector as a whole is being considered or only a quantity of a substance N, changes in resource distribution or concentration lead, according to Sect. 4.1.2, to a change in the state (the entropy) of the system. The decrease in the stock of ecological "advance" that accompanies the extraction of resources is thus also encompassed by the entropy approach. We can therefore establish relationships between the stock of resources, resource extraction, resource concentration and entropy, and thence derive propositions about the temporal development of the energy-(factor-)input into resource extraction.

4.2 The Change in Entropy in the Environmental Sector as a Result of Resource Extraction

We first derive the relationship between the stock of resources, the quantity of resources extracted, and the initial distribution of resources in the environmental sector. As we shall show, the distribution of resource concentration is an appropriate measure for resource distribution. Again we consider resource 1, whose number of mints in a section u of the environment at time $t = 0$ we assume to be $N_1^u(0)$. Let the resource be distributed among I deposits with the capacities $N_1^i(0)$ $(i = 1, \ldots, I)$; thus

$$N_1^u(0) = \sum_{i=1}^{I} N_1^i(0). \tag{4.23}$$

Denoting the number of mints of resource 1 extracted from deposit i in period t by $n_1^i(t)$ we obtain for the number of mints extracted in the periods up to and including period $(s - 1)$

$$\overline{N}_1^u(s-1) = \sum_{t=1}^{s-1} \sum_{i=1}^{I} n_1^i(t) \quad (s = 2, 3, \ldots). \tag{4.24}$$

Assuming that the stocks of resources in section u are changed only by the withdrawal of resources, not by some flow of material from the environment of u or synergetic processes within u, then we obtain for the temporal development of the remaining stock of resources in period s the flow equation

$$\begin{aligned} N_1^u(s) &= N_1^u(0) - \overline{N}_1^u(s-1) = \\ &= \sum_{i=1}^{I} N_1^i(0) - \sum_{t=1}^{s-1}\sum_{i=1}^{I} n_1^i(t) = \\ &= \sum_{i=1}^{I} \left(N_1^i(0) - \sum_{t=1}^{s-1} n_1^i(t)\right) \quad (s = 2, 3, \ldots). \end{aligned} \tag{4.25}$$

If we now divide – as we did in the definition Eq. (4.1) – the stock of resources still remaining in the environment u in period s, $N_1^u(s)$, by the total number of all types of material $N^u(s)$, then we can define a medium resource concentration $\kappa_1^u(s)$ for the material of type 1:

$$\kappa_1^u(s) = \frac{N_1^u(s)}{N^u(s)} \qquad (s = 1, 2, \ldots). \tag{4.26}$$

In the case when waste materials arising in obtaining resources are not rechanneled into section u, we obtain the following relationship between the respective remainder of the total number of all mints $N^u(s)$ and the quantity of resources withdrawn so far, $\overline{N}_1^u(s)$:

$$N^u(s) = N^u(0) - \overline{N}_1^u(s-1). \tag{4.27}$$

The medium resource concentration $\kappa_1^u(s)$ then has the following temporal development:

$$\kappa_1^u(s) = \frac{N_1^u(0) - \overline{N}_1^u(s-1)}{N^u(0) - \overline{N}_1^u(s-1)} = \frac{N_1^u(0) - \sum_{t=1}^{s-1}\sum_{i=1}^{I} n_1^i(t)}{N^u(0) - \sum_{t=1}^{s-1}\sum_{i=1}^{I} n_1^i(t)} \qquad (s = 2, 3, \ldots). \tag{4.28}$$

Differentiation of (4.28) with regard to resource extraction immediately yields that each additional resource extraction $dn_1^i(t)$ at time t lowers the medium resource concentration $\kappa_1^u(s)$. For

$$d\kappa_1^u(s) = \frac{-(N^u(0) - N_1^u(0))}{(N^u(0) - \overline{N}_1^u(s-1))^2}\, dn_1^i(t) < 0 \qquad (4.29)$$

for $N^u(0) > N_1^u(0)$ and $dn_1^i(t) > 0$.

As a consequence of the relationship between entropy and initial concentration derived in Sect. 4.1.2, however, an additional decrease of the resource concentration in the environmental sector is accompanied by an additional increase of entropy in this sector.[10] Any additional utilization of the ecological stock thus entails a decrease of the level of order of the environmental sector and thereby an increase of entropy, as was explained in Chap. 3, Sect. 3.6. This decrease in the level of order, in the form of a reduced medium resource concentration in the deposits, must be compensated in the following periods by an additional input of energy into the separation process; from now on – on average and in the long run – a lower resource concentration prevails in extraction within section u. In the long run, this entails an increase of the mol-specific labor input into resource extraction. But even if the mol-specific labor input remained constant for a certain time during the exploitation of a deposit i (i = 1, ..., I), the latter statement would still retain its validity. A simple graph suffices to show this. Fig. 4.8 represents the temporal development of the exploitation of three deposits. For simplicity we assumed that the three deposits do not differ with regard to the labor input necessary for the appropriation of the base material from which resource 1 is to be separated. It was further assumed that the temporal development of the exploitation of the three deposits is so organized that the deposit with the highest concentration, and thus the lowest mol-specific labor input, is exploited first; it is followed by the deposit with the next highest resource concentration.

[10] In a more detailed investigation – which would, however, considerably complicate our derivations above – this statement only holds if we take into consideration that the resources will be diffusely distributed in the environmental sector after having been used in the economic system. This relationship was already noted above in the context of our second example in Sect. 3.6.

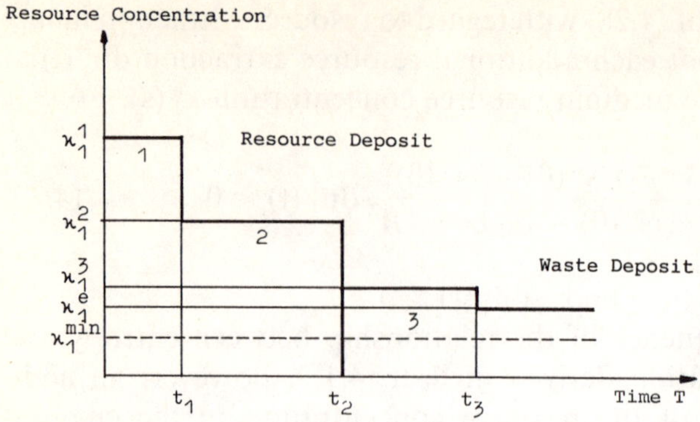

Fig. 4.8. Temporal development of the exploitation of resource deposits with controlled dumping of waste materials

Assuming further that each deposit exhibits a homogeneous resource concentration, this resource concentration and thus the mol-specific labor input into resource extraction does not change during the time intervals $[0, t_1]$, $[t_1, t_2]$, and $[t_2, t_3]$. If at time t_3 there is no longer a deposit whose exploitation is still "worthwhile", then the mol-specific labor input has reached or surpasses an upper bound. This upper bound is determined by a minimum concentration κ_1^{min}, which is dictated by economic considerations. Techniques which depend on this resource will then be eliminated, unless new deposits are found whose concentration surpasses the minimum concentration κ_1^{min} – indicated in Fig. 4.8 by a parallel to the time axis – or precautionary measures are taken.

What precautions, however, can be taken? In order to answer this question we follow the cycle of the use of resource 1 that is needed in production. We begin with the appropriation process of resource extraction, in which the quantity N of a substance is brought from its deposit i with the deposit-specific resource concentration κ_1^i into the production sector. In the second step – the separation process – a portion N_1 is separated which contains nothing but material of type 1 and thus has the resource concentration $\kappa_1 = 1$. Whereas the resources not needed in production are returned to the environmental sector with the remaining quantity $N - N_1$, the separated quantity N_1 of the substance is utilized in its present pure form for the production of capital and consumption goods in the production or in the consumption sector. The material of type 1 leaves the economic system little by little as a part of the emissions. If the emissions contain resource 1 in a concentration κ_1^e, which lies above the minimum concentration κ_1^{min}, then a re-extraction of resource 1 from the emissions may be considered – in Fig. 4.8 at least from time t_3 onwards. The emissions may, for the time being, also be so deposited in a dump in the environmental sector that the resource concentration κ_1^e remains unchanged for as long as possible. Contrary to the diffuse distribution of the resources contained in the emissions over the whole environmental sector, their controlled deposition in a dump creates a reserve which can be tapped later. This precautionary action counteracts the increase in entropy – as does recycling – because the resource concentration in the dump is higher than the one that is obtained if the resources are distributed diffusely over the whole environmental sector.

At this point it becomes clear that the entropy approach is a more encompassing instrument for the investigation of environmental problems than is the material balance approach that was traced to AYRES and KNEESE (1969). Material balances inform us about the whereabouts of types of materials in the respective sectors of the economic and ecological system. Changes in the distribution (or concentration) of the material resulting from extraction and emission, however, are not registered. The entropy approach, on the other hand, enables us to treat the distributional aspects of the use of the environment as a supplier of resources in addition to the quantity aspects. The entropy approach enables us, moreover, to put the different forms of uses of the environment on a common basis and to investigate their interdependencies.

We shall take a closer look at these interdependencies between environmental protection and resource problems later in Chap. 7, where we shall also investigate the precautionary action mentioned above, such as dumping and recycling. In the following third Part of our book, which consists of Chaps. 5 and 6, we shall occupy ourselves exclusively with the use of the environment as a supplier of resources. As a starting point for the integration of the resource problem we shall here use the capital model of BERNHOLZ and FABER (1973), which had also served as the basis for the environment model of NIEMES (1981) introduced in Chap. 2.

4.3 The Limits of Our Approach[11]

Evaluating our approach so far, we believe one should be well aware of the limits of using entropy as an economic variable. We will illustrate this point with the two relationships derived in this part, the pollution function (3.13) and the energy requirement for extraction (4.15). Except perhaps for heat (cf. our remarks at the end of Sect. 3.9), we doubt whether it is possible to specify empirically the functional form of a pollution function (3.13), because of two fundamental difficulties:

a) Scientists often do not yet know what effects a particular pollutant has. Examples of this fact are acid rain and the greenhouse effect. In addition, the interactions between different pollutants have been neglected in the formulation of (3.13), and in particular in the aggregate pollution function (3.14).
b) Even if there is certainty about all these effects in constructing a pollution function, one encounters all the well-known problems of trying to specify an appropriate social welfare function (see. e.g. SEN, 1970:Chap. 8).

Turning to the relationship between energy requirement for extraction and resource concentration, (4.15), we remind the reader that we had to make several heroic assumptions to derive it. This function can therefore only be used as a qualitative representation.

On the positive side, we consider our approach to be helpful in the following respects:

1. Entropy might be used as a new variable to unify the complex subjects of environmental economics and of the economics of exhaustible resources.[12]
2. The thermodynamic approach is a way for economics to get in touch with its own biophysical foundation.
3. Using an entropy approach makes an economist aware of the irreversible nature of the time structure of many environ-

[11] This Section is based on parts of Sect. 3.2 of FABER (1985).

[12] While we have concentrated on exhaustible resources, MUROTA (1984) has made an attempt to develop a framework for renewable resources, in particular with regard to water and forests.

mental and resource processes. This concerns in particular those long- run problems on a macroeconomic level, for which the price-system alone does not provide a sufficient solution.

From the point of view of thermodynamics we can analyze new economic problems, which so far either have yet to be examined or for which we did not have the theoretical framework to do so. These are mainly problems which are on the border-line between the economy and the natural environment. Entropy is a concept which enables us to link these two areas. It makes us aware of problems.

Appendix: Value Charts for Section 4.1.2 (Numerical Tables for the Relationship Between the Mol-Specific Change in Entropy and Resource Concentration)

1. Values for the relationship between the mol-specific change in entropy and the initial concentration

$$S_{N_1} = R\left\{\ln \kappa_1 + \frac{1-\kappa_1}{\kappa_1}\ln(1-\kappa_1)\right\}$$

are listed in Table 4.1:

Table 4.1. Calculation of the mol-specific change in entropy during the extraction process as a function of resource concentration

	A	B	C	D	E
κ_1	$\ln \kappa_1$	$\frac{1-\kappa_1}{\kappa_1}$	$B \ln(1-\kappa_1)$	$A + C$	S_{N_1}
0.01	– 4.6052	99.0000	– 0.9999	– 5.5051	– 46.6030
0.05	– 2.9957	19.0000	– 0.9747	– 3.9704	– 33.0115
0.1	– 2.3026	9.0000	– 0.9486	– 3.2512	– 27.0318
0.2	– 1.6094	4.0000	– 0.8924	– 2.5018	– 20.8010
0.3	– 1.2040	2.3333	– 0.8323	– 2.0363	– 16.9306
0.4	– 0.9163	1.5000	– 0.7662	– 1.6825	– 13.9890
0.5	– 0.6931	1.0000	– 0.6931	– 1.3862	– 11.5254
0.6	– 0.5108	0.6607	– 0.6054	– 1.1162	– 9.2805
0.7	– 0.3567	0.4286	– 0.5160	– 0.8727	– 7.2560
0.8	– 0.2231	0.2500	– 0.4024	– 0.6255	– 5.2007
0.9	– 0.1054	0.1111	– 0.2558	– 0.3612	– 3.0032
0.95	– 0.0513	0.0526	– 0.1576	– 0.2089	– 1.7369
0.99	– 0.0101	0.0101	– 0.0465	– 0.0566	– 0.4706

2. The values of the first derivative of relationship (4.10) with respect to resource concentration

$$d(S_{N_1})/d\kappa_1 = \frac{-R}{\kappa_1^2} \ln(1-\kappa_1)$$

are listed in Table 4.2. We see that the first derivative has its minimum for a resource concentration of approximately 0.7:

Table 4.2. Calculation of the values for the first derivative of relationship (4.10) with respect to resource concentration

κ_1	κ_1^{-2}	$\frac{-R}{\kappa_1^2}$	$\ln(1-\kappa_1)$	$d(S_{N_1})/d\kappa_1$
0.01	10000	− 83144.0	− 0.0101	839.7544
0.05	400	− 3325.76	− 0.0513	170.5892
0.1	100	− 831.440	− 0.1054	87.6338
0.2	25	− 207.860	− 0.2231	46.3736
0.3	11.111	− 92.3813	− 0.3567	32.9524
0.4	6.250	− 51.9650	− 0.5108	26.5437
0.5	4.000	− 33.2576	− 0.6931	23.0508
0.6	2.7778	− 23.0957	− 0.9163	21.1626
0.7	2.0408	− 16.9680	− 1.2040	20.4295
0.8	1.5625	− 12.9913	− 1.6094	20.9082
0.9	1.2346	− 10.2650	− 2.3026	23.6362
0.95	1.1080	− 9.2124	− 2.9957	27.5976
0.99	1.0203	− 8.4832	− 4.6052	39.0668
0.999	1.0020	− 8.3311	− 6.9078	57.5493

Part III

The Use of Scarce Resources with Decreasing Resource Concentration

It was mentioned in the Introduction that we do not want to investigate the changes in the stocks of resources as a result of production and resource extraction only. Corresponding to our thermodynamic interpretation of the problem of resources, we also want to take into consideration the resulting changes of the resource concentration in the environmental sector.[1] We therefore wish to build a bridge between the resource problem on the one hand and the question of environmental damage on the other.

In Chap. 5 we first describe the structure of the resource model. Since it is derived from our initial model, which was discussed in the second Chapter, we now confine ourselves to the explanation of newly added assumptions and features of the model. These are essentially:

1. an added subsector within the production sector, in which a raw material needed for the production of the capital good is separated; and
2. assumptions about the environmental sector, whereby changes in resource quantities and concentrations are considered.

In the preceding Chapter it was shown that the input of factors into resource extraction depends on resource concentration. The formulation of these relationships for appropriation and separation determines the characteristics of our resource model. We emphasize this by the investigation of two model variants. In the first model variant, we abstract from the fact that the appropriation process which precedes the separation process proper also calls for the commitment of factors of production. Only the second model variant will include the utilization of factors of production for both the appropriation and the separation of resources from the environmental sector.

In Chap. 6 we shall derive the conditions which must be fulfilled for an exhaustible resource to be utilized in production. For this purpose we first investigate the economic effects which result from the introduction of a new production process. What we are interested in is the temporal development of the production of

---

[1] Cf. the last Chapter of the present study. Let it be noted that in many studies both of these aspects are not simultaneously investigated. See also the overview by FISHER and PETERSON (1976) and FISHER (1981).

consumption goods, of resource utilization, of resource concentration, and thus also of factor inputs.

For both model variants we shall then derive conditions for the intemporally efficient and optimal use of the environment as a supplier of resources. In particular, it will become apparent, how an increase in resource extraction costs in the long run can be explained within the confines of our model from the properties of the production techniques and ecological conditions.

5 The Integration of the Resource Problem into a Disaggregated Capital Model

5.1 The Structure of the Model

Before turning to the special properties of the new features of the model, let us note once again that in this Chapter we shall not yet deal with the questions of the treatment of waste materials and pollutants. We are thus making:

Assumption 5.1. The emissions which arise in production or consumption and which are channeled into the environmental sector do not cause any impairment of the flow of environmental goods into the economic subsystem.

Therefore we do not consider waste-treatment. We are also excluding from the investigation recycling and the controlled deposition of emissions in the environmental sector in this model:

Assumption 5.2. Resources which are contained in the emissions and which can be utilized in production are not re-extracted from these emissions. Rather, the emissions and the resources contained therein are diffusely distributed over the whole environmental sector.

We are putting special emphasis on these assumptions at this point because we want to highlight the areas of the model to which the problems of both the increasing scarcity of resources and of the treatment of wastes will be combined in Chap. 7. For instance, if a resource is recycled from the emissions for use in production, this activity not only leads to a decrease of resource scarcity, but also under certain conditions to an improvement in the quality of the environment. This will be the case particularly if elements of this resource, which are discharged into the envi-

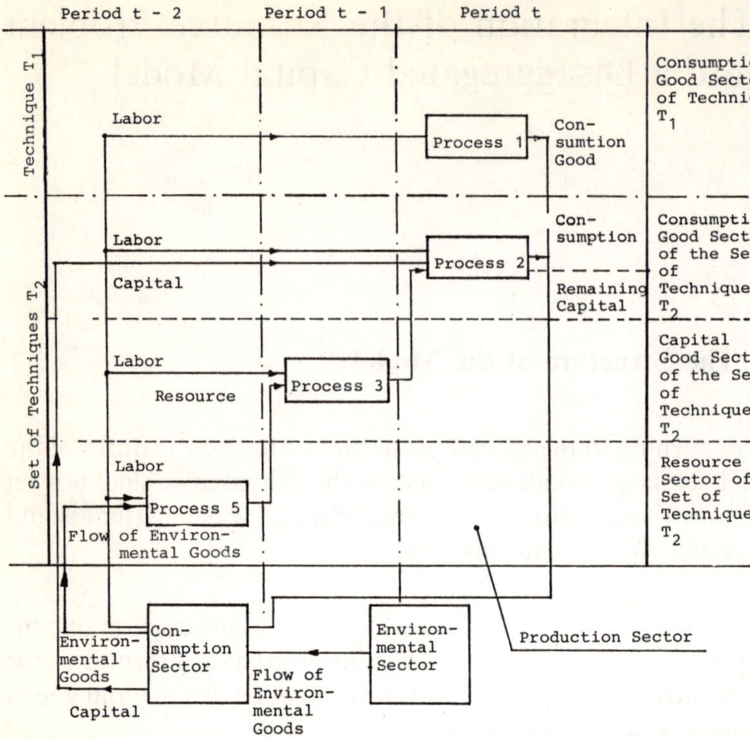

Fig. 5.1. Structure of the resource model

ronment in the form of emissions, disturb the ecological equilibrium and subsequently impair the flow of environmental goods into the economic subsystem.

The structure of the resource model is represented in Fig. 5.1. What has been added to the initial model in Chap. 2, Sect. 2.1.1, is (1) the environmental sector and (2) the resource subsector within the production sector, with the resource appropriation and separation process R_5.[2] In this sector a resource is separated from the type of material contained in a deposit of the environment. This resource is then used for the production of the capital good in the third process. Since according to our assumptions (cf.

[2] The resource appropriation and separation process is given the index 5, since the waste treatment process in the environment-capital model was given the index 4 and we want to stick to this notation.

Chap. 2) the first process – and thus technique $T_1 = (R_1)$ – does not utilize this capital good, only processes 2, 3, and 5, i.e. technique $T_2 = \{R_2, R_3, R_5\}$, are affected by this model extension. The resource model thus derives its peculiar characteristics from technique T_2. For here the special properties of the resource sector on the one hand and the environmental sector on the other, in their function as suppliers of resources, make themselves felt.[3]

5.1.1 The Resource Sector

Having described the main outlines of the structure of the resource model we shall now turn to the resource sector. As was discussed in Chap. 4, the process of resource separation proper in the extraction of resources is preceded by an appropriation process. Material containing the resources needed for production are here transmitted from the environmental sector into the economic subsystem. In the resource sector – as was assumed in our model for the sake of simplicity – a type of material needed for the production of the capital good in process R_3 of the technique $T_2 = \{R_2, R_3, R_5\}$, is separated from these volume elements.

We further assumed (cf. Chap. 4) that only the factor labor is used as an input in the extraction of resources. The relationship between labor input (energy) and resource concentration for the extraction of one unit of the resource which was derived in the fourth Chapter is applied in our model. In this context, two model variants are distinguished

Model Variant 1. Here we assume for each moment of time t, that the labor input into the separation of resources is exclusively a function of resource concentration, of the form (cf. Chap. 4, Sect. 4.2):

$$l_5^I(\kappa(t)) = l^E(\kappa(t)) = \varepsilon(-\ln \kappa(t)), \quad 0 < \kappa(t) < 1. \tag{4.19a}$$

[3] Aspects of uncertainty remain unconsidered in our approach. An extensive analysis of resource extraction under uncertainty can be found in Hoel (1980).

In this model variant we thus begin by abstracting from the fact that the appropriation process in resource extraction also utilizes factors. These are then taken into consideration in the second model variant.

Model Variant 2. Instead of (4.19a) we here assume equation

$$I_5^{II}(\kappa(t)) = l^R(\kappa(t)) = \frac{\alpha}{\kappa(t)} + \varepsilon(-\ln \kappa(t)), 0 < \kappa(t) < 1. \tag{4.21 a}$$

In (4.19 a) and (4.21 a) the input of factors into the fifth production process R_5 depends on the concentration of the resource. In distinction from the other sectors, the resource sector thus encompasses not a single process, but instead a set of processes $\{R_5\}$, whose range of values is determined by the possible resource concentrations. Each of the processes of $\{R_5\}$ thus has the form $R_5(\kappa)$, $0 < \kappa < 1$. In dealing with technique T_2 we are therefore really dealing with a set of techniques $\{T_2\}$, which can replace technique T_1 in the production of one and the same consumption good. For simplicity, we shall in the following maintain the notation T_2.

The input and output coefficients of the processes $R_5(\kappa(t))$ of the resource sector, which correspond to the respective concentrations $\kappa(t)$ at time $t \in (1, \ldots, T)$, are determined as follows. If the levels of activity $x_5(t)$ of the resource processes utilized have a value of 1, then one unit of the resource needed for the production of the capital good is being separated from the other types of materials. As can be seen in Fig. 5.2, the output coefficient of process $R_5(\kappa(t))$ for the separated resource has a value of 1. Assuming that a total (100%) or near total separation from the other types of materials is being achieved, we obtain for any given level of activity $x_5(t)$ precisely $x_5(t)$ units of the desired resource. A number of units of all types of materials corresponding to the reciprocal value of the respective resource concentration has to be extracted from the environmental sector on the input side if one unit of the resource is to be separated. For a resource concentration $\kappa(t) = 0.1$ this is e.g. $1/\kappa(t)$, i.e. 10 units of all types of materials. Depending on the model variant, the input coefficient for the input of labor is determined either by (4.19 a) or by

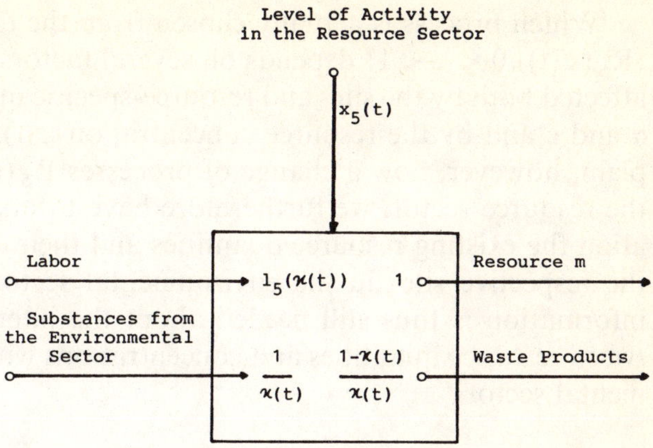

Fig. 5.2. Input- and output-coefficients of the processes in the resource sector for the models of type I and II

(4.21 a). For the extraction of $x_5(t)$ units of the resource a labor quantity of either $x_5(t)\,l_5^{I}(\kappa(t))$ or $x_5(t)\,l_5^{II}(\kappa(t))$ units of labor is needed. Finally, the output coefficient for the types of materials which are channeled into the environment as waste materials from process $R_5(\kappa(t))$ (emissions) has a value of $(1-\kappa(t))/\kappa(t)$. In our example of a resource concentration of 0.1, 9 units of waste materials would thus arise if one unit of the resource needed in production were to be separated from the other types of materials. The number of elements contained in one mol of this substance may here be taken as a unit for the resource.

For the input and output coefficients of a process $R_5(\kappa(t))$ we thus obtain the following relationship[4]:

$R_5(\kappa(t))$: $l_5(\kappa(t))$ units of labor $\oplus$ $(1/\kappa(t))$ units of base material $\rightarrow$ 1 unit of resource $\oplus$ $(1-\kappa(t))/\kappa(t)$ units of waste products.

As usually assumed, input quantities are applied at the beginning of a period while output quantities are available at the beginning of the next period.

[4] Note that a complete separation of a particular commodity is even theoretically impossible. To simplify formal efforts, one unit of raw material should be read as one unit of almost pure raw material.

Which process is actually chosen from the feasible processes $\{R_5(\kappa(t))/0 < \kappa < 1\}$ depends on several factors. The decision is affected both by the site- and resource-specific quantities, such as α and ε and by the resource concentration $\kappa(t)$. In order to explain, however, how a change of processes $R_5(\kappa)$ takes place in the resource sector, we furthermore have to take into consideration the existing resource quantities and their distribution over the respective sites in the environmental sector. In our model information is thus still needed about the interrelationship between resource quantities and concentrations within the environmental sector.

5.1.2 Resource Quantities and Concentrations in the Environmental Sector

Our main aim in demonstrating the relationship between resource quantity, resource concentration, labor input into resource extraction, and the change of entropy in the environmental sector in Chap. 4, Sect. 4.2, was to show that the exploitation of the stock of ecological services within the environmental sector can be slowed down by preventive steps, such as recycling and orderly waste deposition. The relationships derived in Chap. 4, however, can also serve as first indicators as to how the use of the environment as a supplier of resources can be formally represented within our model. The flow equation

$$N_1^u(s) = N_1^u(0) - \overline{N}_1^u(s-1) = \sum_{i=1}^{I} N_1^i(0) - \sum_{t=1}^{s-1} \sum_{i=1}^{I} n_1^i(t) \qquad (s = 1, \ldots, T) \quad (4.25)$$

for instance defines the temporal state of the stock of resources $N_1^u(s)$ of resource 1 in the part u of the environment at time s as a function of the initial stocks $N_1^i(0)$ within the sites $i = 1, \ldots, I$ and of the resource quantities $n_1^i(t)$ extracted up until time $s - 1$. From the equation for the development of the average resource concentration $\kappa_1^u(s)$ at time s we can thus conclude how each extraction of resources from the environmental sector decreases average concentration:

$$\kappa_1^u(s) = \frac{N_1^u(0) - \sum_{t=1}^{s-1} \sum_{i=1}^{I} n_1^i(t)}{N^u(0) - \sum_{t=1}^{s-1} \sum_{i=1}^{I} n_1^i(t)} \quad (s = 1, \ldots, T). \tag{4.28}$$

The extraction of resources thus constitutes an exploitation of the available stock of ecological services. These losses, which we have interpreted as an increase of entropy in the environmental sector, lead to an immediate or subsequent increase of factor input into resource extraction (cf. (4.19) and (4.21)). Relationship (4.28) can thus be interpreted as a link between the environment as a supplier of resources and the economic subsystem. It demonstrates how decisions about the level of extraction $x_5^i(t) = n_1^i(t)$ in periods $t = 1, \ldots, s-1$ determine the total concentration in period s. Thus we have knowledge about the relationship between economic activities and the stock of ecological services still remaining within the environment. For $\kappa_1^u(s)$ is nothing but a measure for the latter. No conclusions can be drawn from (4.28), however, with regard to the sequence of exploitation of the respective resource deposits. For not only the average concentration $\kappa_1^u(s)$ of the resource within the environment but also the concentration within each site has to be taken into consideration in the calculation of an optimal plan. The reason is that, according to (4.19) and (4.21), it determines the labor coefficients in the process of resource extraction in both variants of the model. This means that the sequence of the exploitation of the respective deposits has to be determined simultaneously with the other economic quantities. This would, however, considerably complicate the solution of our problem since we would have to use a procedure of integral programming. We shall thus assume that the sequence of the exploitation of the deposits is already known.[5] As was explained in Chap. 4, Fig. 4.8, it is determined by the level of resource concentration, i.e. the deposit with the highest concen-

[5] HERFINDAHL (1967) has shown that deposits are exploited in a sequence of increasing extraction costs. Cf. GLOBAL 2000 (1981:220). KEMP and LONG (1980), however, have submitted contrary evidence.

tration is exploited first and it is followed by the deposit with the next highest resource concentration.

But even if the sequence of exploitation is determined according to such a rule, a difficulty arises in this context, for the transition from one deposit to another may lead to discrete changes of the resource concentration $\kappa^u(s)$ and thus of the labor coefficients of the resource processes R_5. In order to avoid the mathematical difficulties that arise from such discontinuities (cf. Chap. 6), we shall assume this transition to be continuous. Let the relationship between the concentration $\kappa(s)$ in period s (s = 1, ..., T) and the quantity of the resource extracted up until and including period (s − 1)

$$\bar{R}(s) = \sum_{t=1}^{s-1} x_5(t)^6 \tag{5.1}$$

be described by an exogenously given function:

$$\kappa(\bar{R}(s)) = \kappa(0)\, e^{-\psi \bar{R}(s)} \qquad (s = 1, \ldots, T). \tag{5.2}$$

The value of $\kappa(\bar{R}(s))$ at the beginning of period 1 is

$$\kappa(\bar{R}(0)) = \kappa(0) = \frac{R(0)}{N(0)}.$$

It therefore has the same form and the same value as $\kappa_1^u(0)$ in (4.28). $\kappa(s)$ and $\kappa_1^u(s)$ also fall monotonically with the quantity extracted $\bar{R}(s)$. The positive parameter ψ indicates the gradual decrease of resource concentration. Function (5.2) rapidly decreases for large values of ψ, as can be seen from a comparison of the two graphs in Fig. 5.3. For small values of ψ it decreases slowly. The concentration function (5.2) thus characterizes the environmental sector. Through it the stock of ecological services in the form of the concentration of resources (which nature has given men as an advance, c.f. Sects. 4.1.4 and 4.2) is taken into

[6] Since in the following we shall consider one type of resource only, we shall utilize R(s) instead of $N_1^u(s)$ for the total quantity of the resource and set $\bar{N}_1^u(s) = \bar{R}(s)$.

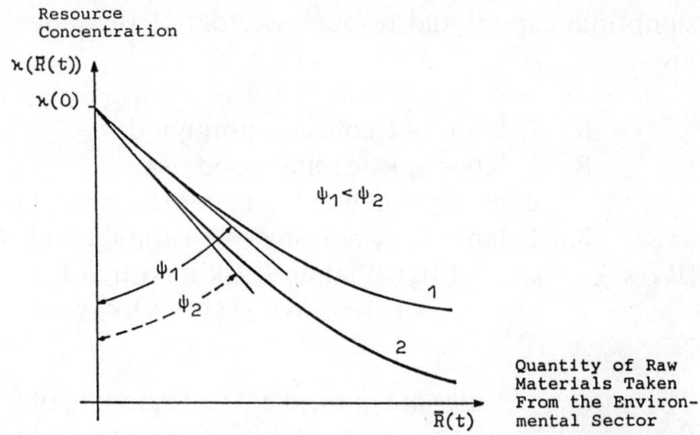

Fig. 5.3. The concentration function in our resource-capital model

consideration.[7] Since this stock of ecological services is different for different resources, the forms of the concentration functions for the different resources differ.

The concentration function (5.2) constitutes an essential extension of our initial model. In the following Sect. 5.2 we shall present the system of constraint equations for our extended model.

5.2 The System of Constraint Equations

We shall first summarize the technology of our resource model. The processes R_1 and R_2 with which the consumption good can be produced have the same form as in Chap. 2, except for the fact that emissions are not considered (c.f. Assumption 5.1). In process R_3, however, r_3 units of the resource are now needed for the production of one unit of the capital good in addition to the l_3 units of labor. The input- and output-coefficients for the con-

[7] The exogenously given resource concentration function can be interpreted differently with respect to the number of deposit sites. It can be a description either of only one deposit site with decreasing resource concentration, or of a number of deposit sites within the environmental sector which are listed according to (5.2) with respect to their concentrations.

sumption, capital and resource sectors of the model are as follows:

$$\begin{aligned}
&R_1: l_1 \text{ labor} \rightarrow 1 \text{ consumption good}\\
&R_2: l_2 \text{ labor} \oplus k_2 \text{ capital good} \rightarrow 1\\
&\qquad \text{consumption good} \oplus (1-c)\,k_2 \text{ capital good}\\
&R_3: l_3 \text{ labor} \oplus r_3 \text{ resource} \rightarrow 1 \text{ capital good}\\
&\{R_5(\kappa)/0<\kappa<1\}: \{l_5(\kappa) \text{ labor} \oplus 1/\kappa \text{ mixture of base material}\\
&\qquad \rightarrow 1 \text{ resource} \oplus (1-\kappa)/\kappa \text{ waste products}\}.
\end{aligned} \tag{5.3}$$

The input quantities are applied at the beginning of a period and the output quantities are available at the beginning of the next period. Technique T_1 consists of process R_1 and technique $\{T_2(\kappa), 0<\kappa<1\}$ of R_2, R_3, and $\{R_5(\kappa), 0<\kappa<1\}$ for all $0<\kappa<1$. (We remind the reader that according to our remarks in Sect. 2.1.1 in processes R_1 and R_2 a flow of environmental goods is needed in addition to the factors labor and capital listed so far. This flow is not mentioned because it is not assumed as scare).

The input coefficient $l_5(\kappa(\bar{R}(s)))$ of the resource sector is either indicated by (4.19 a) or by (4.21 a) – depending on which model variant is being considered. The resource concentration $\kappa(\bar{R}(s))$ as a function of the quantity of resources extracted from the environmental sector until period s, $\bar{R}(s)$, is determined by the concentration function (5.2). The explicit forms of the labor inputs in the resource sector can be obtained by substituting the resource concentration in (4.19 a) and (4.21 a) by (5.2).
For the *first model variant* the labor input in the resource sector is

$$l_5^{I}(\kappa(\bar{R}(s))) = -\varepsilon \ln \kappa(0)\, e^{-\psi \bar{R}(s)} = -\varepsilon \ln \kappa(0) + \varepsilon \psi \bar{R}(s)$$

and for the second model variant

$$\begin{aligned}
l_5^{II}(\kappa(\bar{R}))) &= \frac{\alpha}{\kappa(\bar{R}(s))} + l_5^{I}(\kappa(\bar{R}(s)))\\
&= \frac{\alpha}{\kappa(0)} e^{\psi \bar{R}(s)} + \{-\varepsilon \ln \kappa(0) + \varepsilon \psi \bar{R}(s)\}.
\end{aligned} \tag{5.5}$$

In the first model variant the labor input grows in proportion to[8] the quantity of resources already extracted, $\bar{R}(s)$, while in the second it grows more than in proportion to this. This can be seen in the first two derivatives of the last two expressions:

$$\frac{\partial l_5^I}{\partial \bar{R}(s)} = \varepsilon\psi, \tag{5.6}$$

$$\frac{\partial l_5^{II}}{\partial \bar{R}(s)} = \frac{\alpha\psi}{\kappa(0)} e^{\psi R(s)} + \frac{\partial l_5^I}{\partial \bar{R}(s)} = \frac{\alpha\psi}{\kappa(s)} + \varepsilon\psi. \tag{5.7}$$

In the second model variant the relationship between labor input and resource concentration thus corresponds to the relationship represented in Fig. 4.6 of Chap. 4.

We shall juxtapose both model variants when deriving the optimality conditions in Chap. 6.

Before formulating the restrictions of our model, we shall indicate two assumptions. These are utilized in order to keep the number of variables and of inequalities in our model as small as possible.

Assumption 5.3. Resources cannot be stored. They are fully utilized in production in the period following their extraction.

Under this condition we obtain the following relationship between the levels of activity of the processes in the resource and capital good sectors:

$$x_5(s-1) = r_3 x_3(s). \tag{5.8}$$

This consideration also shows that the duration of the production of the capital good inclusive of that of the resources needed totals two periods. For the production of the consumption good with a technique of $\{T_2(\kappa)/0 < \kappa < 1\}$ we thus need three periods.

In order to avoid having to draw case distinctions we further make:

[8] The resource distribution function (5.2) was exogenous in this model. Since in the first model variant the input of labor grows in proportion to the resource quantity $\bar{R}(s)$, it thus also depends on the resource distribution function chosen.

Assumption 5.4. In all periods of the planning horizon there is a positive quantity of the resource in the environmental sector:

$$R(s) = R(0) - \overline{R}(s) > O \qquad (s = 1,\ldots,T).$$

Since resources are never fully extracted, this condition is usually fulfilled in reality. In our model, it is at least fulfilled for the second model variant because for $\kappa(s) \to 0$ the input of labor needed for the extraction of a further unit of the resource grows beyond all bounds. As a consequence, the extraction of resources must come to an end before resources are fully exhausted.

As in the initial model of Chap. 2, we shall again assume that there is no capital satiation. This condition means that the available quantities of the capital good will in no period suffice to use the total quantity of labor in period s, L(s), in the second process R_2.

Let us now formulate the system of constraint equations of our model. Since this was explained extensively for our initial model in Chap. 2, and since the new parts of the model were already described in the foregoing Section, we now only need to summarize the constraints already known:

$$\begin{aligned}
&\text{(i)} \quad H_Q(s+1) = \sum_{j=1}^{2} x_j(s) - Q(s+1) && \geqq 0\\
&\text{(ii)} \quad H_L(s) = L(s) - \sum_{j=1}^{3} x_j(s)\, l_j - x_5(s)\, l_5(\kappa(\overline{R}(s)) && \geqq 0\\
&\text{(iii)} \quad H_K(s) = K(0) + \sum_{t=2}^{s-1} x_3(t) - c k_2 \sum_{t=1}^{s-1} x_2(t)\\
&\qquad\qquad - x_2(s)\, k_2 && \geqq 0^{9}\\
&\text{(iv)} \quad H_R(s) = x_5(s-1) - x_3(s)\, r_3 && \geqq 0\\
&\text{(v)} \quad H_{\overline{R}}(s) = \overline{R}(s) - \sum_{t=1}^{s-1} x_5(t) && \geqq 0\\
&\text{(vi)} \quad x_j(s) \geqq 0 \qquad (j = 1, 2, 3, 5;\ s = 1, \ldots, T). && (5.9)
\end{aligned}$$

[9] Since it was assumed that no raw materials are available at the beginning of the planning horizon, production of the capital good can only be started in the second period.

As was already mentioned in the Chap. 2, the vectors consisting of the activities of the production processes $x_j(s)$ ($j = 1, 2, 3, 5$ and $s = 1, \ldots, T$) are called admissable production programs π if they fulfill (5.9). We further assume that the quantities of consumption goods produced in period s are consumed in period $s + 1$. Constraints (i), (ii), and (iii) were already explained in a similar form in Chaps. 1 and 2. (i) says that the production of the consumption good $\sum_{j=1}^{2} x_j(s)$ in period s equals or is greater than the demand $Q(s + 1)$ in period $s + 1$. (ii) is the constraint with respect to the factor labor and (iii) is the constraint with respect to the capital good. Constraint (iv) represents the relationship between production of the resource, $x_5(s - 1)$, and the input of the quantity of the resource $r_3 x_3(s)$ in the production of the capital good, as was already indicated in (5.8). (v) is a definition equation for the quantity of the resource extracted from the environmental sector up until period s, $\bar{R}(s)$.

6 Replacement of Techniques over Time

In the preceding Chapter we described the structure of the resource model and found that there were two techniques with which consumption goods can be produced. Technique T_1 consisted of one single production process R_1 only. This process was assumed to produce consumption goods without any non-renewable resources having to be extracted. With the set of techniques $\{T_2(\kappa), 0 \leqq \kappa \leqq 1\}$ the consumption good can only be produced, however, if resources have already been extracted from the environmental sector. It consists of production processes R_2, R_3, and of the set of processes $R_5(\kappa)$, $0 \leqq \kappa \leqq 1$.[1] R_2 serves exclusively the production of consumption goods – as did process R_1. It requires the use of a capital good, however, which must first be produced in process R_3. This in turn requires scarce resources, which have to be extracted from the environment with one of the processes of $\{R_5\}$, depending on the resource concentration. In the context of this resource model, a change from technique T_1 to technique T_2 implies a transition to a production technology which uses non-renewable resources.

In the present Chapter we want to examine how and under what circumstances a transition from technique T_1 to technique T_2 is effected, and to asses the effects of this transition on some of the main economic variables such as consumption, labor input,

[1] Let us remember that $\{R_5\}$ contains an unlimited number of processes since every κ yields a different process $R_5(\kappa)$. As was explained in Chap. 5, Sect. 5.1, we thus have to consider not only one technique $T_2 = \{R_2, R_3, R_5\}$, but an unlimited number of techniques $\{T_2(\kappa)\} = \{R_2, R_3, R_5(\kappa)\}$. For simplicity, we shall in the following often speak of technique T_2 instead of the set of techniques $\{T_2(\kappa), 0 \leqq \kappa \leqq 1\}$.

resource extraction, and relative prices.[2] We start by considering the following questions:

1. In what way can technique T_1 be replaced by technique T_2 in the context of the resource model?
2. What are the effects of a (partial) replacement of technique T_1 by technique T_2 on the production of consumption goods, the extraction of resources from the environmental sector, and the temporal distribution of the concentration of resources?
3. What technological conditions have to be met in order for a replacement to be initiated at all?
4. How long is technique T_2 applied after it has been introduced?
5. Under what conditions is it optimal to apply technique T_2?

In Sect. 6.1 we shall begin by considering the first two questions. We shall describe a transition scheme from technique T_1 to technique T_2 and discuss the effects on the major economic variables. The answers to questions 3 and 4 can be found in Sect. 6.2. There we shall only consider necessary conditions for the application of technique T_2, however. The last question will be dealt with in Sects. 6.3 and 6.4. We shall there derive the optimality conditions for a change of techniques and discuss these in the context of both model variants of Chap. 5.

6.1 The Transition from Technique T_1 to Technique T_2 and Its Effects on the Economic System

6.1.1 A Schema of Replacement Processes in the Resource Model

We now turn to the first question. The possibilities to replace technique T_1 by technique T_2 either partially or totally are directly

---

[2] For further discussions see MAIER (1984). He gives an explicit neo-Austrian analysis of the time profile of production and consumption, if a technology using a non-renewable resource is replaced by a backstop technology. Using a neo-Austrian approach WODOPIA (1986) gives an econometric study on the substitution processes between coal, oil and gas which took place in the electricity sector in the Federal Republic of Germany between 1971 and 1979. As mentioned above FABER, NIEMES and STEPHAN (1983) and STEPHAN (1988) analyze the effects of the introduction of new water treatment plants.

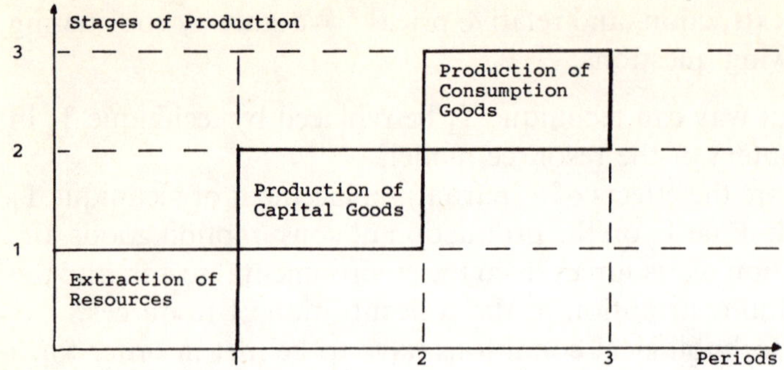

Fig. 6.1. Sequence of production steps and production duration of technique T_2

determined by the structure of the resource model. As was already demonstrated in Sect. 5.1, the three production steps resource extraction, production of the capital good, and production of the consumption good must be applied one after the other if technique T_2 is to be utilized for the production of the consumption good. Since, according to the production assumptions of the model, the production steps can only be applied in this sequence and since each takes one period, the duration of production of one unit of the consumption good with technique T_2 extends over three periods (cf. Fig. 6.1). Included, however, is the duration of production of all intermediate products.

It thus follows from the structure of this model that each replacement process which starts in period t and leads from technique T_1 to technique T_2 follows the following scheme:

1. The production of consumption goods with technique T_1 is reduced in periods t, t + 1, t + 2, and in the following periods.
2. In period t the quantities of the factor labor that are thus set free are used for the extraction of resources with one of the processes of $\{R_5(\kappa), 0 \leqq \kappa \leqq 1\}$.
3. In period t + 1 they are used together with the resources extracted for the production of the capital good in process R_3.
4. In period t + 2 they are used together with the capital goods for the production of the consumption good in process R_2. The first quantities of the consumption good produced are thus forthcoming from technique T_2 at the end of period t + 2 (beginning of period t + 3).

This answers the question in which way technique T_1 can be replaced by technique T_2 in the context of the resource model. Naturally, this process of substitution can extend over more than three periods. As is to be explained in the next Section, it generally will.

6.1.2 The Effects of Replacement Processes in the Resource Model on the Major Economic Variables

In this Section we shall examine our second question: what are the effects over time of a (partial) replacement of technique T_1 by technique T_2 on

a) the distribution of the factor labor between the respective production processes of the model,
b) the production of consumption goods as a whole and the share of both techniques in that whole, and
c) the extraction of resources from the environmental sector and the resulting changes in resource concentration and quantity within that sector.

Suppose that in period t technique T_1 is partially replaced by technique T_2. The effects accompanying the replacement process are already known from the scheme described above. In periods t, t + 1, and t + 2 the production of consumption goods with technique T_1 is reduced, but only at the beginning of period t + 3 are consumption goods forthcoming from technique T_2. Simultaneously, the quantities of labor released by technique T_1 are used in processes R_5, R_3, and R_2 (in that temporal order). Resources are thus extracted from the environmental sector in period t which lowers the average concentration of resources in that sector.

In order to describe the effects that can be observed after an introduction of technique T_2 we shall, for simplicity, start by distinguishing between two different types of development possible.

Development Path A. Technique T_2 is introduced in period t and produces identical amounts of the consumption good in each period starting with period t + 2.

The effects of this form of the production of consumption goods with technique T_2 on the distribution of the factor labor between the techniques and the production of consumption goods with technique T_1 can be explained by three effects.[3]

1. In each period the production of consumption goods with technique T_1 must be at least so reduced that the quantities of the factor labor released suffice to allow the production of consumption goods with technique T_2.
2. In production, the consumption good process R_2 of technique T_2 also uses the factor capital in constant proportions to the factor labor. Since, according to the assumptions of our model, the capital good deteriorates in production at rate c, a corresponding quantity of the capital good must be produced in each period in order to be able to maintain the production of consumption goods with technique T_2 at a given level. This implies a further reduction of the production of consumption goods with technique T_1, for quantities of labor must be made available for process R_3.
3. Capital goods can only be produced, however, if resources have been extracted from the environmental sector one period earlier (cf. Sect. 5.1). This implies a third reduction of the production of consumption goods with technique T_1 in each period. The quantities of labor necessary for the extraction of resources are equally taken from technique T_1.[4]

In the development path A, resources must thus be extracted from the environment in every period. Every resource extraction, however, is accompanied by a reduction in the average resource concentration in the environmental sector (cf. the concentration function (5.2)). Constant resource extraction thus implies that from period to period resources with decreasing resource concentration have to be extracted. This temporal change of resource concentration in the environmental sector can be read from Fig. 6.2. The more resources are extracted from the environment in the course of time, the lower is the concentration of the remainder.

---

[3] Let it be remembered that within this model efficient production is possible only if the factor labor is fully employed (cf. Sect. 5.2).

[4] As a result of the assumption that there is no capital satiation (cf. Sect. 5.2) the quantity of labor used in T_1 suffices for this purpose.

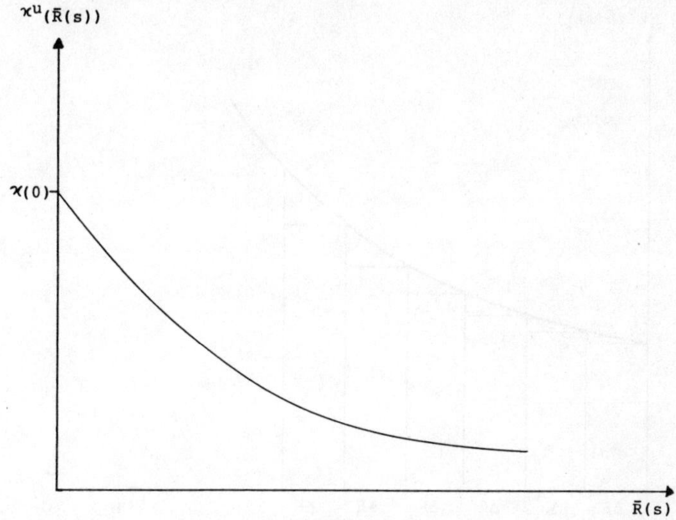

Fig. 6.2. Average concentration of resources in the environmental sector as a function of the quantities extracted

Since resource concentration and labor input into extraction are inversely related (cf. Chap. 4), the development path A is accompanied by an increase in the input of labor into resource extraction. This relationship is depicted in Fig. 6.3. If we now take into consideration that during the development along path A identical quantities of resources (ΔR) are extracted in each period, then we can elucidate the relationship described above with the help of Fig. 6.3.

Let us return from these investigations to the third effect which results from a (partial) transition to technique T_2. Since the input of labor into resource extraction has to be increased from period to period it follows that over the course of time more and more factors are withdrawn from the production of consumption goods with technique T_1. We thus obtain the development of the production of consumption goods with techniques T_1 and T_2 along the development path A that is schematically depicted in Fig. 6.4.

Development Path B. Technique T_2 is introduced in period t, but the deterioration of the capital good is not replaced. This implies

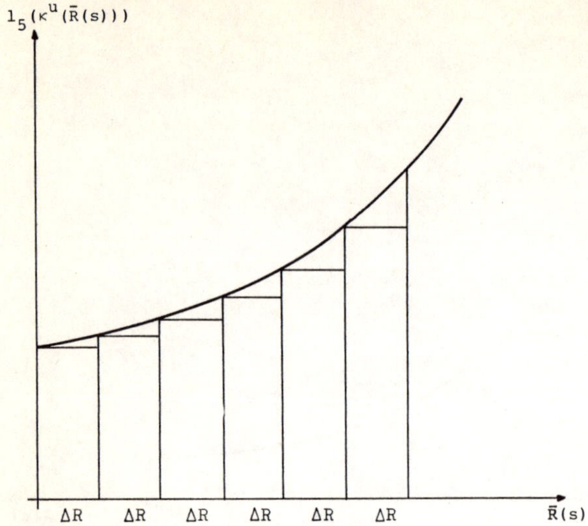

Fig. 6.3. Labor input in the resource sector $l_5(\kappa^u(\overline{R}(s)))$ as a function of the quantity of resources N extracted from the environment

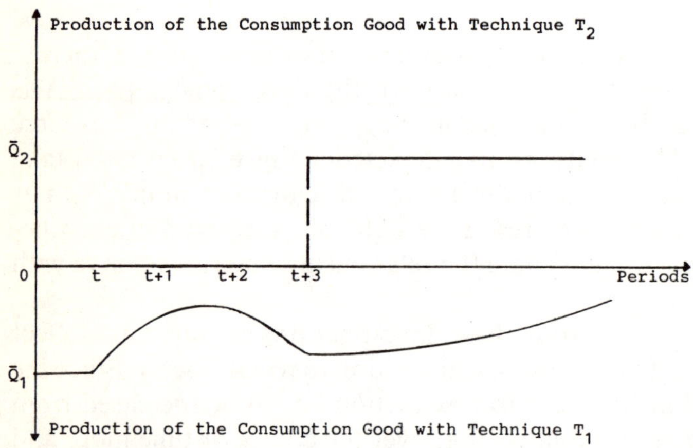

Fig. 6.4. Production of consumption goods with technique T_1 and technique T_2 along development path A

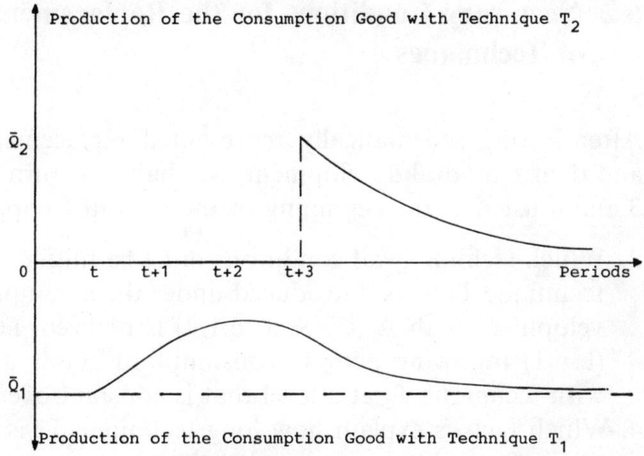

Fig. 6.5. Production of consumption goods with technique T_1 and technique T_2 along development path B

that only in the phase of replacement of technique T_1 by technique T_2, i.e. in periods t and t + 1, are resources extracted and capital goods produced. Thereafter, neither resources are extracted nor capital goods produced. The effects of this form of the production of consumption goods over time is thus explained by one effect only. The production of consumption goods with technique T_1 is reduced so that the quantities of the factor labor released suffice for the production of capital goods with process R_2. However, as a result of the deterioration of the capital good and the ensuing reduction of the production of the consumption good with technique T_2, the production of the consumption good with technique T_1 increases with time until it has reached its original level (cf. Fig. 6.5).

Development paths A and B, however, not only describe two special forms of future development accompanying the replacement of technique T_1 by technique T_2. Since we are considering a linear model, every other feasible development path which accompanies a replacement of technique T_1 by technique T_2 can be represented as a linear combination of paths A and B at suitable points of time. We have thus described, in principle, all possible replacement paths.

6.2 Necessary Conditions for the Replacement of Techniques

After having schematically represented replacement processes and their temporal development, we shall now turn to questions 3 and 4 listed at the beginning of the present Chapter:[5]

3. Which technological conditions must be fulfilled in order for technique T_2 to be introduced under the assumption that development path A (cf. Sect. 6.1.2) is realized, i.e. that in all (t + 1) following periods consumption goods are produced with technique T_2 at a level that is constant over time?
4. Which factors explain how long technique T_2 is actually applied after it has been introduced?

Since for the moment we are only interested in the technological conditions, we make:

Assumption 6.1. The objective is the efficient production of consumption goods.

Since technique T_1 produces consumption goods with labor alone, it follows from Assumption 6.1: under the objective that identical quantities of the consumption good be produced in periods $s > t + 1$, technique T_2 is introduced in period t only if the following is true. In each period $s > t + 1$ the input of a certain quantity of labor to technique T_2 yields at least the same quantity of the consumption good as does the input of the same amount of labor to technique T_1. Considering the linear production structure of our model, this postulate means: the sum of the labor quantities that are used in technique T_2 directly for the produc-

---

[5] With respect to questions 3 and 4 listed at the beginning of this Chapter, only the analysis of development path A is of interest. For development path B the answers to these questions follow directly from the definition. For the sake of completeness these are listed in the following:
To question 3): Since in the develoment along path B after the introduction of technique T_2, i.e. from period $t + 2$ onwards, only process R_2 is being applied, it follows from the condition of efficient production, process R_2, and thus technique T_2, is only used along path B if $l_1 > l_2$.
To question 4): Since in the development along path B the deterioration of the capital good is not replaced, it follows that over the course of time technique T_2 is resubstituted by technique T_1.

tion of one unit of the consumption good and indirectly for the maintenance of the output level in the following periods must be no larger than the amount of labor that is needed for the production of one unit of the consumption good with technique T_1.

We have thus derived a necessary condition. It must be met if consumption goods are to be produced with technique T_2 under Assumption 6.1. If we now try to express this condition in terms of the production coefficients of the production technology of our model, we obtain for every period s the relationship

$$l_1 \geqq l_2 + l_3 \, ck_2 + l_5 (\kappa (s-1)) \, r_3 \, ck_2 \, . \tag{6.1}$$

On the left-hand side of inequality (6.1) l_1 represents the amount of labor necessary to be able to produce one unit of the consumption good with technique T_1. On the right-hand side we find the sum of the direct and indirect amounts of labor necessary to produce one unit of the consumption good in period s under the condition: in the following period it must likewise be possible to produce one unit of the consumption good with technique T_2. We shall now interpret the right hand side of inequation (6.1) in detail:

1. l_2 indicates the amount of labor needed directly in process R_2 for the production of one unit of the consumption good.
2. $l_3 \, ck_2$ describes the amounts of labor necessary in order to replace the quantity of the capital good which has been lost through deterioration in the consumption process R_2. For if one intends to maintain the production of the consumption good with technique T_2 at a constant level (development path A), one must replace the quantity of the capital good that has been used up.
3. For the production of the capital good, however, one also needs resources, which must be extracted from the environmental sector. The labor requirement in resource extraction is in turn a function of the resource concentration in the environmental sector (cf. Chap. 4 and Fig. 6.3). If the production of consumption goods with technique T_2 is to be maintained in the future, then not only the quantity of the capital good that has been used up, ck_2, has to be replaced; in addition, resources to the amount of $r_3 \, ck_2$ must have been extracted (one

period earlier). The labor that is here required is indicated by the expression $l_5(\kappa(s-1))\, r_3\, ck_2$.

With inequality (6.1) a necessary condition, in technological terms, has been derived for the application of technique T_2 with efficient production. We can thus follow in reverse. If condition (6.1) is no longer met in period s, then with efficient production, technique T_2 will be discontinued in one of the following periods. We have thus answered the question of the duration of the application of technique T_2.

From our considerations in Chaps. 4 and 5 we know that for every period s

$$l_5(\kappa(s-1)) \leqq l_5(\kappa(s)) \tag{6.2}$$

and thus

$$\begin{aligned} &l_2 + l_3\, ck_2 + l_5(\kappa(s-1))\, r_3\, ck_2 \\ &\quad \leqq l_2 + l_3\, ck_2 + l_5(\kappa(s))\, r_3\, ck_2 . \end{aligned} \tag{6.3}$$

Insofar as consumption goods are to be produced with technique T_2 in every period (development path A), resources must be extracted in every period; $\kappa(s)$ thus decreases in the course of time while $l_5(\kappa)$ increases (cf. (5.4) and (5.5)). Therefore there exists a period t* for which

$$l_1 \leqq l_2 + l_3\, ck_2 + l_5(\kappa(t^*-1))\, r_3\, ck_2 . \tag{6.4}$$

With efficient production (Assumption 6.1), no further investment takes place from this period onwards and from period t* less and less of the consumption good is produced with process R_2. The amounts of labor released by the discontinuation of R_3 and R_5 and the reduction of R_2 are used again in process R_1.

Period t* depends on the quantity of the consumption good produced with T_2. The larger this quantity, the more resources must be extracted, the faster thus the decrease of $\kappa(s)$, and the sooner period t* is reached as a result of the rise of the labor coefficient $l_5(\kappa(s))$.

The quantity of the consumption good produced in total in every period is determined also by the demand – in addition to the technological conditions and the production factors.

So far the demand has remained largely unconsidered in this Chapter; only the production side in the use of scarce resources with decreasing resource concentration has been investigated. Thus no welfare effects have been compared and no conditions for the actual choice of a production program have been derived. In the following two Sections we shall now include the demand side. With the help of the shadow prices for the consumption good and the production factors we shall discuss whether, and for how long, it is advantageous to produce with technique T_2 in addition to technique T_1.

6.3 Optimality Conditions[6]

For the derivation of optimality conditions we additionally make

Assumption 6.2. The objective is to maximize the (quasi-concave) intertemporal welfare function

$$W(Q(2), \dots, Q(T)) = \sum_{\tau=2}^{T} (1+\delta)^{1-t} W_t(Q_t) . \qquad (6.5)$$

Function (6.5) differs from the intertemporal welfare function discussed and utilized in Chap. 2 only in that the environmental good U is now not an argument in the function. This simplification has been introduced because we are here concerned with the question of an optimal use of resources for the production of consumption goods only. In Chap. 7, where we shall investigate environmental protection and resource problems simultaneously, we shall abandon this simplification.

A feasible production program is called optimal if it maximizes the objective function (6.5). According to the procedure of KUHN-TUCKER this means (cf. e.g. HADLEY 1964:185–212): if

[6] Instead of employing a finite horizon framework, we could have used an infinite horizon approach. The general results would not change. The mathematics, however, would be much more complicated. Hence, in order to keep the mathemetical effort as small as possible, we work with finite horizon models. For a comparison between finite and infinite horizon approaches and the related theoretical difficulties see STEPHAN (1985, 1986).

from the welfare function (6.5) and the system of constraint Eqs. (5.9), Sect. 5.2, we form the LAGRANGEan function

$$V = \sum_{t=2}^{T} (1+\delta)^{1-t} W_t(Q_t) + \sum_{t=1}^{T} \{\sum_{h} p_h(t)\, H_h(t)\},$$

where for the simplification of the notation Q, L, K, R and $\bar{R}$ are replaced by h, then every optimal production program must fulfill the following conditions:[7]

(i) $$\partial V/\partial Q(s+1) = (1+\delta)^{-s}\, \partial W_{s+1}/\partial Q(s+1) - p_Q(s+1) \begin{cases} = 0 & \text{for } Q(s+1) > 0 \\ \leqq 0 & \text{for } Q(s+1) = 0 \end{cases}$$

(ii) $$\partial V/\partial x_1(s) = p_Q(s+1) - l_1\, p_L(s) \begin{cases} = 0 & \text{for } x_1(s) > 0 \\ \leqq 0 & \text{for } x_1(s) = 0 \end{cases}$$

(iii) $$\partial V/\partial x_2(s) = p_Q(s+1) - l_2\, p_L(s) - k_2\, p_K(s) - ck_2 \sum_{t=s+1}^{T} p_K(t) \begin{cases} = 0 & \text{for } x_2(s) > 0 \\ \leqq 0 & \text{for } x_2(s) = 0 \end{cases}$$
$$(s = 1, \dots T-1)$$

(iv) $$\partial V/\partial x_3(s) = \sum_{t=s+1}^{T} p_K(t) - l_3\, p_L(s) - r_3\, p_R(s) \begin{cases} = 0 & \text{for } x_3(s) > 0 \\ \leqq 0 & \text{for } x_3(s) = 0 \end{cases}$$
$$(s = 1, \dots, T)$$

(v) $$\partial V/\partial x_5(s) = p_R(s+1) - l_5(\kappa(\bar{R}(s)))\, p_L(s) - \sum_{t=s+1}^{T} p_{\bar{R}}(t) \begin{cases} = 0 & \text{for } x_5(s) > 0 \\ \leqq 0 & \text{for } x_5(s) = 0 \end{cases}$$
$$(s = 1, \dots T-1)$$

(vi) $$\partial V/\partial \bar{R}(s) = p_{\bar{R}}(s) - x_5(s) \frac{\partial l_5(\kappa(\bar{R}(s)))}{\partial \bar{R}(s)} \cdot p_L(s) \begin{cases} = 0 & \text{for } \bar{R}(s) > 0 \\ \leqq 0 & \text{for } \bar{R}(s) = 0 \end{cases}$$
$$(s = 1, \dots, T)$$

---

[7] The constraints $H_h(t)$ were given in (5.9).

(vii) $H_h(s) = 0$ for $p_h(s) \geqq 0$
$H_h(s) > 0$ for $p_h(s) = 0$
$p_Q(s+1), p_R(s+1) \geqq 0$ $(s = 1, \ldots T-1)$
$p_L(s), p_K(s), p_{\overline{R}}(s) \geqq 0$ $(s = 1, \ldots, T)$.
(6.6)

The LAGRANGEan multipliers $p_h(s)$ can be interpreted as shadow prices of the variables $h = Q, L, K, R, \overline{R}$.

As was seen in Sect. 6.2, we cannot assume that both techniques T_1 and T_2 are applied in all periods. In order to simplify the formal derivation of the optimality conditions for the use of non-renewable resources, however, we make

Assumption 6.3. [8] With optimal production the planning horizon [1, T] consists of a time span $[t_1, t_2]$, $3 \leqq t_1 < t_2 \leqq T-2$, in which techniques T_1 and T_2 are simultaneously used for the production of consumption goods and in which all three processes R_2, R_3, and R_5 of T_2 are applied.

We shall now interpret the system of optimality conditions for both variants of our resource model. We shall confine ourselves to the time span $[t_1, t_2]$ in which, according to the Assumption 6.3, both techniques are being applied.

6.4 Interpreting the Optimality Conditions

6.4.1 Model Variant I

According to Assumption 6.3 conditions (i) to (vii) of the system (6.6) are equations if period s lies within the interval $[t_1, t_2]$. This means in particular:

1. For the consumption good we obtain the following shadow price which is determined by the demand side

$$p_Q(s+1) = (1+\delta)^{-s}\, \partial W_{s+1}/\partial Q(s+1),$$
$$(s \in \{t_1, \ldots, t_2\}). \quad (6.7)$$

[8] Conditions for the truth of Assumption 6.3 were derived in (6.1).

At the optimum, the shadow price for the consumption good must correspond to its social marginal cost, which is determined by differentiation of the intertemporal welfare function. Two different effects are thus taken into consideration, for $\partial W_{s+1}/\partial Q\,(s+1)$ expresses the preference for consumption with respect to period $s+1$, while $(1+\delta)^{-s}$ expresses the preference for consumption over time. Thus both preferences with respect to the period in question and intertemporal preferences are taken into consideration.

2. From (ii) and (iii) we obtain the following conditions. For the production of one unit of the consumption good with the process R_1 we get

$$p_Q(s+1) = l_1\, p_L(s) \tag{6.8}$$

and for production with process R_2 using the capital good

$$p_Q(s+1) = l_2\, p_L(s) + k_2\, p_K(s) + ck_2 \sum_{t=s+1}^{T} p_K(t)\,. \tag{6.9}$$

These two optimality conditions indicate that the shadow price of one unit of the consumption good must correspond to the marginal cost[9] of the consumption good. They are thus also called no-profit-conditions. Whereas process R_1 only involves marginal costs for the factor labor, viz. $l_1\, p_L(s)$, process R_2 involves marginal costs for the capital needed, too. These consist of two components:

a) the costs $k_2\, p_K(s)$ for the services of the capital good which are required per unit of the consumption good produced;

b) the opportunity costs which are caused by the fact that as a result of the deterioration of the capital good by ck_2 this quantity is no longer available in the following periods, i.e.

$$ck_2 \sum_{t=s+1}^{T} p_K(t)\,.$$

[9] It is a consequence of the linearity of the technology used that here and in the following marginal cost is equal to average cost.

As was already explained in the second Chapter, we can determine from these considerations the shadow price for one unit of the capital good as

$$\bar{p}_K(s) = \sum_{t=s+1}^{T} p_K(t). \tag{2.22}$$

3. We shall in the following use this short form for the shadow price of one unit of the capital good. From (iv) we then obtain the no-profit-condition for the capital good process R_3 as

$$\bar{p}_K(s) = l_3\, p_L(s) + r_3\, p_R(s), \quad t_1 \leqq s \leqq t_2. \tag{6.10}$$

$\bar{p}_K(s)$ is not only determined by the marginal costs $l_3\, p_L(s)$ for the factor labor, but also by the marginal costs for the resources, which are needed for the production of the capital good and were extracted in the preceding period. Of special interest in our resource model is what the constituents are of the marginal costs of the resources which are used as an intermediate product in the production of the capital good. We shall thus interpret these costs in detail.

4. Before the resource can be applied in the capital good process R_3 of technique T_2, it must have been both appropriated from the environment in the previous period and separated. From (v) we obtain for period $s\,(t_1 \leqq s \leqq t_2)$[10] the no-profit-condition

$$p_R(s) = l_5(\kappa(\bar{R}(s-1)))\, p_L(s-1) + \sum_{t=s}^{T} p_{\bar{R}}(t). \tag{6.11}$$

The shadow price for one unit of the resource, $p_R(s)$, is determined by the sum of two terms. The first indicates the marginal costs for the labor input, which arise in period $s-1$ in the extraction of one unit of the resource made available for period s. As in Chap. 1, Sect. 1.2.1.1, the second term can be interpreted as a royalty. For its form is similar to that of the opportunity costs which represent the foregone uses, $q(t)$ in

[10] Note that compared with (v) from (6.6) we have here stepped back one period from $s+1$ to s in order to be able later to directly substitute expression (6.10) into (6.11).

Eq. (1.4). In order to show that the second term likewise indicates these consequential costs, we must first explain, however, the significance of the shadow prices $p_{\bar{R}}(t)$, $s \leqq t \leqq T$, which constitute the sum $\sum_{s}^{T} p_{\bar{R}}(t)$.

For this purpose, let us consider (vi) and (vii) in the optimality conditions (6.6):

$$0 \leqq p_{\bar{R}}(s) \leqq x_5(s) \frac{\partial l_5(\kappa(\bar{R}(s)))}{\partial \bar{R}(s)} p_L(s).$$

In all periods s, in which resources are extracted, $x_5(s) > 0$ and thus $\bar{R}(s) > 0$; therefore we can substitute the equality sign:

$$p_{\bar{R}}(s) = x_5(s) \frac{\partial l_5(\kappa(\bar{R}(s)))}{\partial \bar{R}(s)} p_L(s). \tag{6.12}$$

In those periods s, however, in which resources are no longer extracted, $x_5(s) = 0$, and thus, according to (6.6) (vii)

$$p_{\bar{R}}(s) = 0, \tag{6.12a}$$

where, according to (6.6) (vi) and (6.12a), Eq. (6.12) holds in all cases. If in model variant I we now substitute $\partial l_5(\kappa(\bar{R}(s)))/\partial \bar{R}(s)$ by Eq. (5.6) we obtain for (6.12) the specification

$$p^{I}_{\bar{R}}(s) = x_5(s)\, \varepsilon\psi\, p_L(s). \tag{6.13}$$

If we also take into consideration that for model variant I the labor input in the resource sector is explicitly described by (5.4), then we obtain from (6.11) by substitution of (6.13) and (5.4) the shadow price of one unit of the resource as

$$p^{I}_{R}(s) = \{-\varepsilon \ln \kappa(0) + \varepsilon\psi \bar{R}(s-1)\}\, p_L(s-1) + \varepsilon\psi \sum_{t=s}^{T} x_5(t)\, p_L(t). \tag{6.14}$$

From the first term on the right side in (6.14) we can see that the labor input in resource extraction in period s is the higher the larger are the quantities of resources that were extracted from the environmental sector in the previous periods. These total $\overline{R}(s-1)$. Only for the borderline case that the resource distribution parameter ψ is zero does the labor input in resource extraction remain unchanged. For in this case the resource concentration in the environmental sector remains on the level of the initial concentration $\kappa(0)$ for as long as there remain any resources at all. If the resource distribution parameter ψ is greater than zero, however, then the labor input in resource extraction depends on the quantity of resources that were previously extracted. Here an analogy makes itself felt between the use of the environment as a supplier of resources and as a recipient of pollutants. Let us remember Sect. 2.2.1.2. There it was shown that for damage functions with increasing marginal damage the shadow price of the immissions,[11] $p_S(s)$, depends also on the prior damage that was caused by emissions $S_1^n, \ldots, S_{s-1}^n$ previously discharged (cf. 2.1.2.1).

Every extraction of resources leads to a decrease of the resource concentration in the environmental sector. The environment does not reverse this decrease of its own accord. The stock of ecological "advances" (we may also call it ecological capital) is thus exploited. Such exploitation can be interpreted as an increase of entropy in the environment in accordance with the relationships established in Chap. 4. That this exploitation of ecological advances has its effects also in the succeeding periods is affirmed by Eq. (6.14). As long as the resource distribution parameter ψ is not equal to zero, all resource extraction causes subsequent costs which arise as a result of the increase of the labor input in resource extraction (a consequence of the decrease of the resource concentration in the environmental sector). As we are still to see in Chap. 7, this increase can be limited by preventive actions such as recycling or the controlled deposition of waste materials with a certain resource concentration. The size of these consequential costs, which can also be interpreted as intertemporal external effects

[11] We remind the reader that the emission are the pollutants which leave a production process, while the immissions are that part of emission which remains in the environment after the diffusion process has taken place (cf. Sect. 2.1.2.1 how the immissions change from period to period).

of the use of the environment as a supplier of resources, depends not only on ψ, but especially on the sum $\sum_{t=s}^{T} x_5(t)\, p_L(t)$. As already noted above, here again is an analogy to the use of the environment as a recipient of pollutants. In the second Chapter, we showed that the shadow price (or the marginal damage) of the net emissions is equal to the sum of the effluent damage that is caused by the net emissions in periods s to T.

5. With the help of the shadow price thus determined, $p_R^I(s)$, we are now in the position to indicate the shadow price for the capital good, $\bar{p}_K^I$. If we substitute (6.14) into (6.10), then we obtain for period s, $t_1 \leqq s \leqq t_2$, and for Model variant I the following marginal costs:

$$\begin{aligned}\bar{p}_K^I(s) = {} & l_3\, p_L(s) \\ & + r_3\{[-\varepsilon \ln \kappa(0) + \varepsilon\psi \bar{R}(s-1)]\, p_L(s-1) \\ & + \varepsilon\psi \sum_{t=s}^{T} x_5(t)\, p_L(t)\}. \end{aligned} \tag{6.15}$$

From these we can also determine the shadow price $p_K^I(s)$ for the services of one unit of the capital good per period.[12] There is a simple relationship between the latter and the shadow prices for a unit of the capital good for two succeeding periods, s and s + 1:

$$\bar{p}_K(s-1) - \bar{p}_K(s) = \sum_{t=s}^{T} p_K(t) - \sum_{t=s+1}^{T} p_K(t) = p_K(s). \tag{6.16}$$

If we take into consideration that for all periods s the quantity of resources extracted is $\bar{R}(s-1) = \bar{R}(s-2) + x_5(s-1)$, then we obtain for all periods s, $t_1 \leqq s \leqq t_2$, the expression

$$\begin{aligned}p_K^I(s) = {} & l_3[p_L(s-1) - p_L(s)] \\ & + r_3[-\varepsilon \ln \kappa(0) + \varepsilon\psi \bar{R}(s-2)] \\ & \cdot [p_L(s-2) - p_L(s-1)]. \end{aligned} \tag{6.17}$$

---

[12] If one wants to determine the total marginal costs for the use of one unit of the capital good per period, then the marginal costs for the deterioration, $c\bar{p}_K(s)$, must be added to $p_K(s)$.

6. In order to determine the marginal costs for the production of one unit of the consumption good with technique T_2 for the first model variant, we need only substitute the two shadow prices $\bar{p}_K^I(s)$ and $p_K^I(s)$ into the expression (6.9). Then we obtain for $t_1 \leqq s \leqq t_2$ [13]

$$
\begin{aligned}
p_Q^I(s+1) = {} & l_2 p_L(s) + k_2 \{l_3 (p_L(s-1) - p_L(s)) \\
& + r_3[-\varepsilon \ln \kappa(0) + \varepsilon\psi \bar{R}(s-2)] \\
& \cdot [p_L(s-2) - p_L(s-1)]\} \\
& + ck_2 \{l_3 p_L(s) + r_3 \{[-\varepsilon \ln \kappa(0) \\
& + \varepsilon\psi \bar{R}(s-1)]\, p_L(s-1) \\
& + \varepsilon\psi \sum_{t=s}^{T} x_5(t)\, p_L(t)\}\}\,. \qquad (6.18)
\end{aligned}
$$

In the first line of (6.18) the first term on the right side indicates the direct marginal costs for the labor input in the process R_2 of technique T_2, while in the second term we find the marginal costs for the services of k_2 units of the capital good. The marginal costs of the deterioration of ck_2 units of the capital good are listed in the third and forth line. In addition to those partial results that were derived in detail above, and summarized in the above expression, (6.18) contains some further important insights. The rate of deterioration c influences the consequential costs via the resource extraction. If c is low, then the effects of the last partial term of (6.18) are also low. This term represents the consequential costs of the extraction of resources. A low rate of deterioration will also entail that – ceteris paribus – replacement investment will not be made to the same extent. For, with a given quantity of resources that has been transformed into a capital good, production can now be kept up for a longer period. This in turn entails that the exploitation of the ecological advances proceeds at a slower pace. Technique T_2 can thus be used for a longer period.

6.4.2 Model Variant II

In model variant I we abstracted from the fact that the appropriation process which precedes the separation process also

[13] Let it be remembered that according to Assumption 6.3 technique T_2 is applied in the time interval $[t_1, t_2]$ only.

requires factors of production. These are now taken into consideration in model variant II. We shall see that in this case the labor input into resource extraction increases at a higher rate than is the case in the first model variant. For with decreasing resource concentration more mints than before have to be appropriated from the environment in order to obtain one unit of the resource. In variant II the set of techniques $\{T_2\}$ will thus tend to be discontinued earlier, because the increasing scarcity of certain resources will have much more serious effects than in the case of the first model variant.[14] This effect is thus at the center of the derivation of the shadow prices for the second model variant. We shall therefore concentrate on these differences between both model variants and refrain from a detailed presentation of the shadow prices. Since expressions (6.7) and (6.12) are valid for the first and the second model variant alike, a renewed discussions of these is superfluous here. Since the following formal derivations are rather cumbersome we have summarized them in the following appendix.

Appendix: Derivation of Shadow Prices of Model Variant II

1. We begin with the determination of the shadow price of one unit of the resource, $p_R^{II}(s)$. For this purpose we first determine the shadow prices $p_{\bar{R}}^{II}(t)$ of the resource quantities $\bar{R}(t)$ already extracted from the environmental sector. We thus replace the partial derivative of the labor input with respect to the resource quantity extracted from the environmental sector up until s, $\bar{R}(s)$, by (5.7). Taking into consideration (6.13) we then obtain for $t_1 \leqq s \leqq T$ instead of (6.12)

$$\begin{aligned} p_{\bar{R}}^{II}(s) &= \frac{\alpha\psi}{\kappa(s)} x_5(s)\, p_L(s) + p_{\bar{R}}^{I}(s) \\ &= \frac{\alpha\psi}{\kappa(0)} e^{\psi\bar{R}(s)} x_5(s)\, p_L(s) + p_{\bar{R}}^{I}(s). \end{aligned} \tag{6.19}$$

[14] This relationship can be formally recognized if one substitutes (5.4) and (5.5) into (6.1).

We see that a reduction of the resource concentration caused by the extraction of resources leads not only to an increase in the shadow price $p_R^I(s)$, but also to an additional, now more than proportional increase in the shadow price $p_R^{II}(s)$. This confirms the conjecture, expressed at the beginning of this Chapter, that the decrease of the resource concentration has a stronger effect on the marginal costs of technique T_2 in the second model variant than in the first.

2. This is even more obvious for the shadow price $p_R^{II}(s)$, because it contains the sum of the marginal costs of resource extraction. The substitution of (6.19) into (6.11) and consideration of (5.5), (6.14), and (6.13) yields

$$\begin{aligned} p_R^{II}(s) &= p_R^I(s) + \frac{\alpha}{\kappa(s)} p_L(s-1) \\ &\quad + \alpha\psi \sum_{t=s}^{T} \frac{1}{\kappa(t)} x_5(t)\, p_L(t) \\ &= p_R^I(s) + \frac{\alpha}{\kappa(0)} \{e^{\psi\bar{R}(S)} p_L(s-1) \\ &\quad + \psi \sum_{t=s}^{T} e^{\psi\bar{R}(t)} x_5(t)\, p_L(t)\}. \end{aligned} \tag{6.20}$$

Compared to the first model variant two new terms have been added. These represent

a) the marginal costs, $\alpha p_L(s-1)/\kappa(s)$, of the appropriation process, in which one unit of the resource is won; and
b) the marginal costs of the negative effect that the decrease of the resource concentration has on the appropriation process in the succeeding periods s to T.

The consequential costs of appropriation arise for the following reason. Due to the reduction of resource concentration in period s, more must be extracted from the environmental sector in the following periods in order for the same amount of the resource to be separated. In this context it is appropriate to summarize the consequential costs of resource extraction that arise in the separation process, as indicated in (6.14), with those that arise in the appropriation process. Utilizing the symbols $\tilde{p}_R^I(s)$ or $\tilde{p}_R^{II}(s)$, according to the model variant under consideration, for these consequential costs (or value of the intertemporal external effects) we obtain the following values for the total consequential costs incurred

$$\tilde{p}_R(s) = \varepsilon\psi \sum_{t=s}^{T} x_5(t)\, p_L(t) \tag{6.21}$$

$$\tilde{p}_R^{II}(s) = \varepsilon\psi \sum_{t=s}^{T} x_5(t)\, p_L(t) + \alpha\psi \sum_{t=s}^{T} \frac{x_5(t)}{\kappa(t)} p_L(t); \tag{6.22}$$

instead of (6.22) we can also write

$$\tilde{p}_R^{II}(s) = \psi \sum_{t=s}^{T} \left[\varepsilon + \frac{\alpha}{\kappa(t)} \right] x_5(t)\, p_L(t). \tag{6.23}$$

We can see from this expression that the consequential costs in the second model variant, $\tilde{p}_R^{II}$, are higher than in the first. The reason is that the decrease of the resource concentration not only impedes the separation, but also the appropriation process in resource extraction in the following periods. Formally, the two model variants thus differ in that the quantity α is zero in the first model variant and positive in the second.[15]

3. With the help of the shadow price of the resource which is used as a factor of production in the manufacture of the capital good, as determined in (6.20), the shadow price for the services of the capital good can be indicated for the second model variant. Substituting (6.20) into (6.10) and taking into consideration (6.15) we obtain the expression

$$\begin{aligned} \bar{p}_K^{II}(s) = \bar{p}_K^{I}(s) \\ + r_3 \left\{ \frac{\alpha}{\kappa(s)} p_L(s-1) + \alpha\psi \sum_{t=s}^{T} \frac{1}{\kappa(t)} x_5(t)\, p_L(t) \right\} \end{aligned} \tag{6.24}$$

[15] As we know from the fourth Chapter, α is a deposit site specific variable that was assumed constant in our model. The deposit sites were specified by us with the help of their resource concentration. It would be possible to differentiate the sites with the help of the value of α. An example for this is the resource model of BECKMANN (1982); here the transport costs of the resources to their point of processing were used as a differentiating characteristic.

In the second model variant, the marginal costs of the production of one unit of the capital good exceed the shadow price $\bar{p}_K^I$ by

a) the indirect marginal costs $r_3 \alpha p_L(s-1)/\kappa(s)$ of the appropriation of those units in the resource sector from which the amount of resources needed is won; and
b) the indirect marginal costs thus caused in the following periods s to T, in the form of a consequential impediment of the appropriation process in resource extraction,

$$r_3 \alpha \psi \sum_{t=s}^{T} x_5(t)\, p_L(t)/\kappa(t).$$

If the production of the capital good is particularly resource intensive, i.e. r_3 is particularly large, then these indirect marginal costs increase in importance. Measures aimed at decreasing of the input coefficient r_3 thus decrease the influence of these indirect marginal costs.

4. This relationship also holds for the shadow price of the services of one unit of the capital good per period, $p_K^{II}(s)$, since it is simply an implication of the above expression, as was already shown in its determination for the first model variant in (6.17). As in (6.17) we obtain $p_K^{II}(s)$ from the difference

$$\begin{aligned} p_K^{II}(s) = \bar{p}_K^{II}(s-1) - \bar{p}_K^{II}(s) &= p_K^I(s) \\ &+ r_3 \left\{ \frac{\alpha}{\kappa(s-1)} p_L(s-2) - \frac{\alpha}{\kappa(s)} p_L(s-1) \right. \\ &\left. + \alpha\psi \frac{x_5(s-1)}{\kappa(s-1)} p_L(s-1) \right\}. \end{aligned} \tag{6.25}$$

Since $\kappa(s) = \kappa(s-1)\, e^{-\psi x_5(s-1)}$, (6.25) yields

$$\begin{aligned} p_K^{II}(s) &= p_K^I(s) + \frac{r_3 \alpha}{\kappa(s-1)} \\ &\cdot \left\{ p_L(s-2) - p_L(s-1) \frac{1 - \psi x_5(s-1)\, e^{-\psi x_5(s-1)}}{e^{-\psi x_5(s-1)}} \right\}. \end{aligned} \tag{6.26}$$

For small values of the activity level $x_5(s-1)$, this can also be approximated by

$$p_K^{II}(s) \approx p_K^{I}(s) + \frac{r_3 \alpha}{\kappa(s-1)} \{p_L(s-2) - p_L(s-1)\}; \tag{6.27}$$

for in this case we obtain – as a first approximation – for the following two expressions of the last term in the brackets of (6.26) the values

$$e^{-\psi x_5(s-1)} \approx 1 \quad \text{or} \quad \psi x_5(s-1) e^{-\psi x_5(s-1)} \approx 0.$$

This approximated value in (6.27) likewise confirms that in the second model variant the shadow price of the services of one unit of the capital good per period exhibits a stronger increase than is the case in the first model variant. This is due to the more difficult appropriation in resource extraction; with falling resource concentration more mols have to be appropriated from the environmental sector in the second model variant than in the first.

5. With the latter shadow prices for the capital good, $\bar{p}_K^{II}(s)$ and $p_K^{II}(s)$, we can finally also indicate the marginal costs for the production of one unit of the consumption good with technique T_2 for the second model variant. Substituting (6.24) and (6.27) into the optimality condition (6.9), valid for both model variants alike, we obtain for these marginal costs, i.e. for the shadow price of the consumption good, the approximative value

$$p_Q^{II}(s) \approx p_Q^{I}(s) + \frac{k_2 r_3 \alpha}{\kappa(s-1)} [p_L(s-2) - p_L(s-1)] + c k_2 r_3 \alpha \cdot \left\{\frac{1}{\kappa(s)} p_L(s-1) + \psi \sum_{t=s}^{T} \frac{1}{\kappa(t)} x_5(t) p_L(t)\right\}. \tag{6.28}$$

The resource concentrations that can be found in the denominators of the last two expressions respectively confirm once more that consideration of the appropriation process causes the marginal costs for the production of the consumption good with technique T_2 to increase faster. Initial cost advantages of technique T_2 with respect to technique T_1 will thus – ceteris paribus – be levelled earlier if not only the separation process but also the appropriation process requires a factor input.

Part IV

Environmental Protection and Resources

7 Interdependencies Between Environmental Protection and Resource Use Exemplified for Recycling and Deposition

After having discussed aspects of the resource problem within our capital theoretic approach in the preceding two Chapters, we shall now investigate two interdependencies between environmental protection and the resource problem. These are recycling (the recovery of resources from waste materials) and deposition (the controlled dumping of waste materials in the environmental sector).

For a better understanding of the present Chapter we wish to recall the approach in this book. In the first part we gave an overview of dynamic environmental models from an economic viewpoint. There we concentrated on aspects of environmental protection. Resource problems were, however, also taken into consideration in the last two Sections of the first Chapter. In the second Chapter we then developed our own environmnental model.

An approach to natural science, and thus ecological explanations of environmental and resource problems can be found in the second Part. Above all we found ourselves able to utilize these considerations in the derivation of relationships between resource quantity, resource concentration, entropy change, and energy- and factor input.

In the third Part we formulated our resource model in accordance with these relationships. One important result of our analysis was the finding that the increase in extraction costs over the course of time is caused by an increase in the input of energy needed for resource extraction. This derives from the fact that the utilization of scarce resource quantities with decreasing resource concentration leads to an increase in entropy within the environmental sector. Changes in the economic sector are thus explained by phenomena from natural science.

In the present Chapter we now intend to combine the "environmental model" (of Chap. 2 of Part I) with the "resource model" (of Part III) into an overall model. We shall illustrate, for the cases of recycling and deposition, that a vantage point has thus been created, from which interdependencies between environmental protection and resource use can be analyzed.

In order to be able to analyze these relationships in Sect. 7.1 we shall combine the environmental model of Chap. 2 and the resource model of Chap. 5 into one comprehensive model. In Sect. 7.2 we shall then use this model to investigate recycling and deposition. In Sect. 7.3 we shall discuss how one can avoid wasting low entropy and in Sect. 7.4 we give an outlook.

7.1 Optimal Resource Use and Environmental Protection in an Environment-Resource Model

7.1.1 The Model

Since the structures of the environment-capital model and of the resource model have already been extensively explained,[1] we can now confine ourselves to the description of the integration of these two models.

In Chap. 5, Assumption 5.1, it was assumed for simplicity that the emissions which are caused in production or consumption, and which are returned to the environmental sector, do not impair the quality of the environment.

This assumption must now be relaxed if we wish to study the interrelationships between environmental and resource problems. Here, as in Chap. 2, we therefore take into consideration that emissions are a burden on the environment, and that the quality of the environment is valued by the households. The intertemporal welfare function thus has the same form as in Chap. 2. Not only the quantity of the consumption good Q (t) but also the

[1] Cf. e.g. Figs. 2.1 and 5.1, Sects. 2.1.1 and 5.1, and the two systems of constraint equations in the appendix to Chap. 2 and in Sect. 5.2.

flow of environmental goods U (t) enters the welfare function as an argument, i.e.

$$W = \sum_{t=2}^{T} (1 + \delta)^{1-t} W_t(Q(t), U(t)), \tag{2.14}$$

where the properties listed in (2.15) apply.

After these preliminary remarks we can combine the environment-capital model of Chap. 2 with the resource model of Chap. 5, to give a comprehensive model. Formally, this is achieved by extending the technology of the resource model described in Chap. 5 by the waste treatment process R_4 of Chap. 2. The structure of this model is depicted in Fig. 7.1. With its help the relationship between the resource sector (the set of processes $\{R_5\}$), the capital good sector (R_3), and the consumption good sector (R_1 and R_2) can be recognized. First, resources are extracted in the resource sector with the input of labor. These are then used in the production of the capital good. As a rule, this causes a change in the stock of the capital good. It grows if net investment is positive in the respective period. With optimal production, the stock of the capital good is apportioned – in accordance with the respective weights in the welfare function – in each period to the production of the consumption good and the treatment of emissions which arise in the production of consumption goods.

Let us now investigate under what conditions environmental protection and resource use are combined with one another. This is the case, for instance, if resources are needed in waste treatment, for the capital goods for waste treatment are produced with resources. There is thus a conflict of objectives between waste treatment, i.e. environmental protection, and the use of scarce resources with decreasing resource concentration. An increase in waste treatment activities in the long-run leads to a decrease in net emissions, immissions, and thus to an increase of the quality of the flow of environmental goods on the one hand. On the other, it necessitates an increased resource use, thereby lowering the average resource concentration in the environmental sector and thus increasing entropy. In the absence of compensation of such an increase in entropy by some actions and processes in the environmental sector, the factor input into resource extraction in-

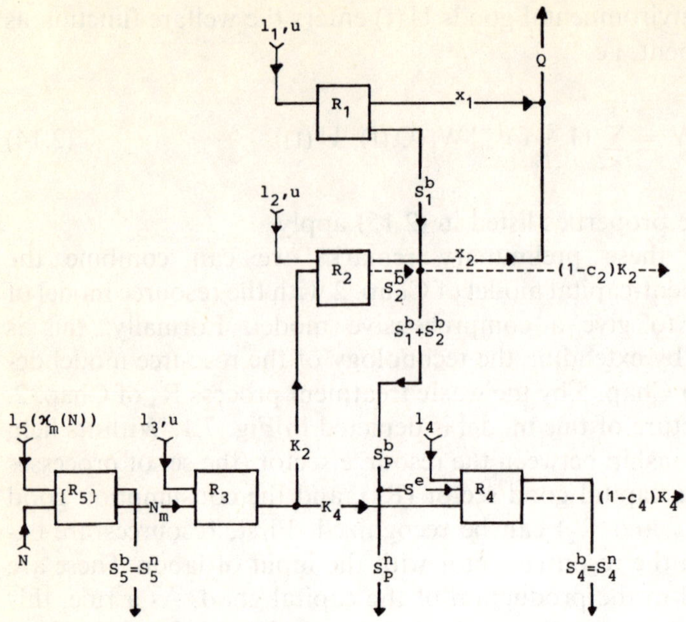

x_j = Quantity of the Consumption Good from Process R_j (j=1,2)
Q = Quantity of the Consumption Good
l_j = Factor Labor in Process R_j (j=1,2,3,4,5)
u = Non-Scarce Environmental Goods
N = Particles of Types m=1,...,M from the Environmental Sector
N_m = Particles of Type M
K_j = Quantity of the Capital Good Used in Process R_j (j=2,4,)
s_j^b = Gross Emissions in Process R_j (j=1,2,4,5)
s_p^b = Gross Emissions of Production
s^e = Emissions to be Treated
c_j = Rate of Deterioration (j=2,4)

Fig. 7.1. Model structure of the "environment-capital model" with capital formation in the waste treatment sector

creases with time according to Eqs. (4.19) or (4.21).[2] The advantages of waste treatment are thus accompanied by the disadvantages of an increased factor input in resource extraction. Environmental protection and resource use must thus be mutually co-ordinated. In the following we shall discuss criteria for such co-ordination with the help of optimality conditions.

---

[2] In the following we shall no longer differentiate between these two versions. The quantity of labor in the resource processes $\{R_5(\kappa), 0 \leqq \kappa \leqq 1\}$ will simply be denoted by $l_5(\kappa)$.

7.1.2 Optimality Conditions

Since the model that is here to be discussed is a summary of the models of Chaps. 2 and 5, most of the optimality conditions are identical with those that were already discussed in Chaps. 2 and 6. We shall therefore confine ourselves here to the discussion of those optimality conditions that relate to the relationships between environmental protection actions and resource use.

As already explained, the link between the increasing scarcity of resources on the one hand, and environmental degradation on the other, is the production of the capital good. For waste treatment requires capital goods, which are produced with resources. The shadow price of the capital good, $\bar{p}_K(s)$, has been determined in Chap. 6 as

$$\bar{p}_K(s) = l_3 p_L(s) + r_3 p_R(s). \tag{6.10}$$

The first term on the right hand side here represents labor costs, while the second indicates resource costs for the production of one unit of the capital good.

The costs $p_R(s)$ for the production of one unit of the resource have also already been determined in Chap. 6 as

$$p_R(s) = l_5(\kappa(\overline{R}(s-1)))\, p_L(s-1) + \sum_{t=s}^{T} p_{\overline{R}}(t). \tag{6.11}$$

Insertion of the shadow price of the resource $p_R(s)$ from (6.11) into (6.10) yields

$$\bar{p}_K(s) = l_3 p_L(s) + r_3 \{l_5(\kappa(\overline{R}(s-1)))\, p_L(s-1) + \sum_{t=s}^{T} p_{\overline{R}}(t)\}. \tag{7.1}$$

The first term here indicates the costs for the direct labor input in the production of one unit of the capital good, whereas the second term indicates the costs for the quantity of the resources thus required. These consist of:

1. the costs for the labor input in resource extraction in the preceding period, $r_3 l_5(\kappa(\overline{R}(s-1)))\, p_L(s-1)$; and

2. the consequential costs $r_3 \sum_{t=s}^{T} p_{\bar{R}}(t)$, which were caused by the resource extraction in period s – 1 for periods s to T.

We now turn to the costs of waste treatment. As was shown in Chap. 2, the shadow price for one unit of emissions treated, $p_{S^e}(s)$, is determined according to equation

$$p_{S^e}(s) = l_4 \, p_L(s) + k_4 \, p_K(s) + c_4 \, k_4 \sum_{t=s+1}^{T} p_K(t) . \qquad (2.21)$$

The first term contains the labor costs while the second and third terms contain the costs for the use of the capital good.

Taking into consideration at this point (cf. Chap. 2) that the following relationship holds between the price of one unit of the capital good in period s, $\bar{p}_K(s)$, and the price for its services in the following periods

$$\bar{p}_K(s) = \sum_{t=s+1}^{T} p_K(t) , \qquad (2.22)$$

we obtain from (7.1) and (2.21)

$$\begin{aligned} p_{S^e}(s) = {} & l_4 \, p_L(s) + k_4 \, p_K(s) + c_4 \, k_4 \, l_3 \, p_L(s) \\ & + c_4 \, k_4 \, r_3 \, l_5 (\kappa (\bar{R}(s-1))) \, p_L(s-1) \\ & + c_4 \, k_4 \, r_3 \sum_{t=s}^{T} p_{\bar{R}}(t) . \end{aligned} \qquad (7.2)$$

This expression shows how the waste treatment costs depend on the costs of the resource use in waste treatment. The lower the resource concentration $\kappa(\bar{R}(s-1))$ the higher are:

1. the labor input $l_5(\kappa(\bar{R}(s-1)))$ and thus the waste treatment costs; and
2. the intertemporal consequential costs $\sum_{t=s}^{T} p_{\bar{R}}(t)$ of resource use. (Cf. the interpretation of this sum in Sect. 6.1 above following Eq. (6.11).)

The costs of resource use are on the other hand also influenced by the intertemporal consequential costs that arise from the fact that net emissions are returned to the environmental sector.

Their shadow price $p_{S^n}(s)$ is equal to the sum of the future immission damages caused by a unit of net emissions that is returned to the environmental sector in period s (cf. the explanation given in Sect. 2.1.2), i.e.

$$p_{S^n}(s) = \sum_{t=s}^{T} (1+\delta)^{-t} W'_U(t+1)(-G'_S(t))\,\Phi^{t+1-s}. \quad (2.25)$$

This shadow price is determined both by social and by ecological variables i.e. by the social discount rate δ and the marginal welfare $W'_U(t+1)$ $(t = s, \ldots, T-1)$ on the one hand, and by the marginal damage $-G'_S(t)$ of immissions $S(t)$ $(t = s, \ldots, T-1)$ and the diffusion factor Φ on the other.

From Chap. 2 we know that at the optimum the shadow price of one unit of net emissions corresponds to the waste treatment costs per unit, i.e.

$$p_{S^n}(s) = p_{S^e}(s). \quad (2.21)$$

Due to (2.21) we can equate expressions (2.25) and (7.2). From this we recognize how the shadow price of net emissions depends on resource concentration.

We now want to investigate how the resource use in production and waste treatment affects the price of one unit of the consumption good at the optimum. For this purpose let us remember that – taking into consideration (2.22) – optimal production with technique T_2 requires that

$$p_Q(s+1) = l_2\, p_L(s) + s_2\, p_{S^b}(s) + k_2\, p_K(s) + c_2\, k_2\, \bar{p}_K(s). \quad (2.19)$$

At the optimum the price per unit of the consumption good is thus determined by the labor costs, $l_2\, p_L(s)$, the costs for the treatment of gross emissions, $s_2\, p_{S^b}(s)$, as well as by the costs for the capital good used inclusive of its deterioration, i.e. $k_2\, p_K(s) + c_2\, k_2\, \bar{p}_K(s)$. It was also shown in Chap. 2 that at the optimum the price of gross emissions, $p_{S^b}(s)$, corresponds to that of the pollutants treated, $p_{S^e}(s)$. Thus we obtain from (2.19) by insertion of (7.1) and (7.2)

$$\begin{aligned} p_Q(s+1) = {} & l_2 p_L(s) + k_2 p_K(s) + s_2 \{l_4 p_L(s) \\ & + k_4 p_K(s) + c_4 k_4 l_3 p_L(s) \\ & + c_4 k_4 r_3 l_5 (\kappa(\bar{R}(s-1))) p_L(s-1) \\ & + c_4 k_4 r_3 \sum_{t=s}^{T} p_{\bar{R}}(t)\} + c_2 k_2 \{l_3 p_L(s) \\ & + r_3 [l_5 (\kappa(\bar{R}(s-1))) p_L(s-1) \\ & + \sum_{t=s}^{T} p_{\bar{R}}(t)]\}\,. \end{aligned} \tag{7.3}$$

As we see, the last term of (7.3) contains the costs that arise in the production of one unit of the consumption good with process R_2 as a result of the capital input and the resource use accompanying it. Since emissions arise during the production of the consumption good, these cause the consequential costs of waste treatment indicated in the first term of the fourth line. We have thus demonstrated that in the production of the consumption good with technique T_2, costs result in all succeeding periods from the indirect input of resources in waste treatment and production.

Note especially that the opportunity costs of today's resource use, in the form of a decreasing resource concentration in the future, are taken into consideration in the optimal commodity prices.

In the following Section we shall be dealing with two different aspects of the resource and environmental problem, viz. recycling and deposition.

7.2 Interdependencies Between Environmental Protection and Resource Use with Recycling and Deposition

Recycling increases in importance – if so far only in certain parts of the economy. For automobile scrapping, for instance, SCHENKEL (1980:172) found that

> "One automobile contains, amongst other materials and on average, 760.0 kg iron and steel, 32.5 kg aluminum, 7.9 kg copper, 26.6 kg lead, 10.2 kg zinc, and 0.6 kg tin; of these 740.0 kg iron waste, 9.4 kg aluminum,

3.1 kg copper, 22.6 kg lead, 4.1 kg zinc, and 2.9 kg of diverse materials are recovered." (Our translation).

In spite of relatively high rates of recovery in certain areas (cf. MOLL 1978:218 ff.), SCHENKEL (1980:172) estimates the contribution of recycling to the alleviation of the resource problem to be, on the whole, very small:

> "... Recycling without restrictions on the increase in the use of native raw materials is only an insignificant contribution to the conservation of resources. Only the prolongation of the length of use and a slowing down of the increase of production in combination with recycling brings an actual release." (Our translation).

Under contemporary conditions it will thus only make sense to engage in recycling in certain cases.

Let us note in passing that an amelioration of the waste-political framework can contribute to an increase in the recovery of resources. For today this framework often stands in the way of recycling (cf. POHLE 1981). It is being proposed, for example, from various sides that the use of one-way packaging should be restricted by legal regulations, if not even totally forbidden in certain areas (cf. MOLL 1978:204–5).[3]

7.2.1 Recycling in the Environment-Resource Model

We shall now treat aspects of recycling in the context of the environment-resource model presented in Sect. 7.1. For this purpose we make:

Assumption 7.1. The pollutants and waste materials arising in the production of consumption goods also contain resources that are necessary for the production of capital goods.

This assumption can be justified as follows. As before we assume that only the resource in question is scarce, and that it is used exclusively for the production of the capital good in process R_3. It is known from the description of the model that the capital good is used in the production of the consumption good in pro-

[3] This type of regulation of the allocation differs fom that considered by us in our investigation, since we make exclusive use of the price mechanism as an economic regulative mechanism.

cess R_2 and in waste treatment in process R_4. Since the capital stocks employed there deteriorate with rates of deterioration c_2 and c_4 respectively, the emissions $S_2^b(s)$ and $S_4^b(s)$ arising in processes R_2 and R_4 contain the resource. Let us assume, as we did in Chap. 2, a centralized waste treatment. The emissions $S_1^b(s)$ of process R_1 which do not contain the resource are then mixed with the emissions $S_2^b(s)$ of process R_2. A portion of these emissions is treated in process R_4. The rest, i.e. the net emissions of production $S_P^n(s) = S_P^b(s) - S^e(s)$, is returned to the environmental sector. The waste materials or "gross emissions of waste treatment", $S_4^b(s)$, are likewise returned to the environmental sector. These contain quantities of resource which result firstly from the deterioration of the capital stock that has been accumulated and secondly which are determined by the amount of waste materials there treated, $S^e(s)$. (Cf. Fig. 7.2; the initial material for the recycling process is here denoted by S^r.)
According to these considerations the initial material from which resources are to be recovered by recycling is determined by:

1. the centrally included gross emissions of production, $S_P^b(s)$; and
2. the waste materials or gross emissions of waste treatment, $S_4^b(s)$.

In the following we shall only investigate the latter of these two possibilities.

For this purpose the technology of the waste treatment sector is extended by a set of recycling processes $\{R_6(\kappa)\}$, with which resources can be extracted from the gross emissions of the waste treatment sector. In this model there are thus two competing types of resource extraction possible; resource extraction from the environment and recycling.

We now assume that the set of recycling processes $\{R_6(\kappa)\}$ possesses the same technological characteristics as the extraction processes $\{R_5(\kappa)\}$. This means that we make:

Assumption 7.2. Recycling only requires labor as a factor and the factor input is a function of the concentration of the resource $\kappa(S_4^b(s))$ in the emissions $S_4^b(s)$.

Since recycling and extraction yield the same resource we can further make:

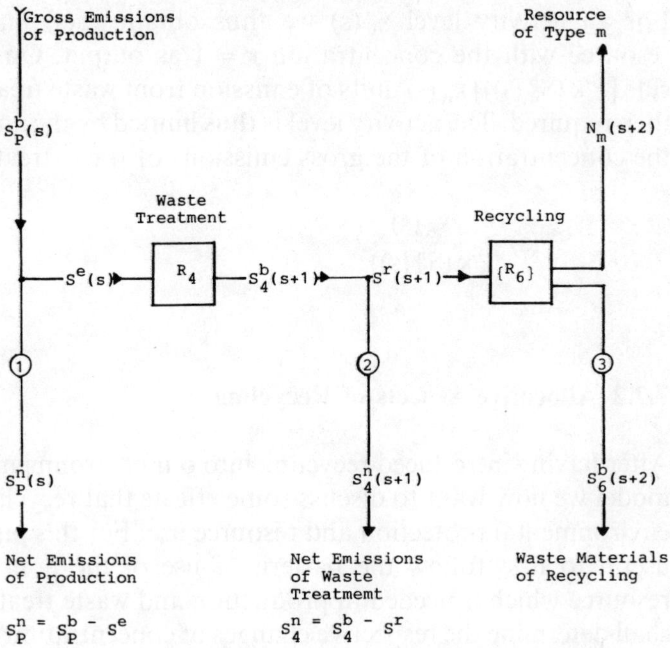

Fig. 7.2. Net emission flows with central waste treatment and recycling

Assumption 7.3. The input- and output-coefficients of the recycling processes $\{R_6(\kappa)\}$ are identical to those of the extraction processes $R_5(\kappa)$ (model variant I).

For each process of the set of processes $\{R_6(\kappa)\}$, the same special functional relationship between labor input and resource concentration applies as in the extraction process, i.e.

$$l_6(\kappa(S_4^b(s+1))) = -\varepsilon \ln \kappa(S_4^b(s+1)). \tag{7.4}$$

If we assume, as we did in extraction, that an almost perfect (100%) separation of the desired resource from other substances is effected, we can describe recycling as follows:

$$R_6(\kappa(t)):\ l_6(\kappa(S_4^b(s))) \text{ units of labor} \oplus 1/\kappa(S_4^b(s)) \text{ units of gross emmissions} \rightarrow 1 \text{ unit of the resource} \oplus \frac{1-\kappa(S_4^b(s+1))}{\kappa(S_4^b(s))} \text{ units emission of recycling.}$$

For an activity level $x_6(s)$ we thus obtain $x_6(s)$ units of the resource with the concentration $\kappa = 1$ as output. On the input side $[1/\kappa(S_4^b(s))]\,x_6(s)$ units of emission from waste treatment are then required. The activity level is thus limited by the amount and the concentration of the gross emissions of waste treatment, i.e.

$$S_4^b(s) \geqq \frac{x_6(s)}{\kappa(S_4^b(s))}. \tag{7.5}$$

7.2.2 Allocative Aspects of Recycling

After having introduced recycling into our environment-resource model we now want to discuss some effects that recycling has on environmental protection and resource use. For this purpose it is useful to first follow the pattern of use of the non-renewable resource which is needed in production and waste treatment. We shall determine the respective changes of concentration and waste treatment at each level of use.

A quantity of substances N with a resource concentration $\kappa(s) < 1$ is appropriated at time s from a deposit in the environmental sector. Extraction in the production sector increases the concentration to $\kappa(s+1) = 1$. This is accompanied by a simultaneous increase in the level of order in the production sector. The entropy of the production sector decreases. On the other hand, the resource extraction is accompanied by an increase in entropy in the environment, as was explained in detail in Sect. 4.2 of the fourth Chapter.

If one now intends to preserve the resulting increase of the level of order in the production sector, one ought not to mix the resource with other substances in the production of goods. Already in the production of the capital good, however, the first "dilution" occurs. For in addition to the factor labor, many other free environmental goods which are needed in conjunction with the scarce factors of production are here – as in other processes – also employed. One unit of the capital good K thus contains the exhaustible resource only in a concentration of

$$\kappa(K(s+2)) < \kappa(s+1) \leqq 1. \tag{7.6}$$

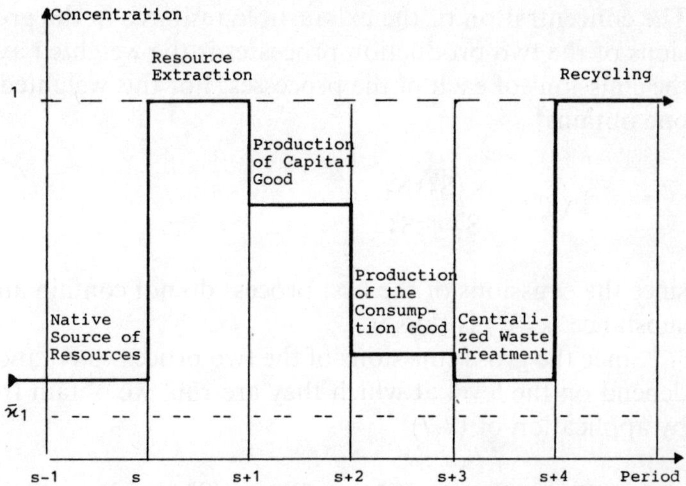

Fig. 7.3. Development of the resource concentration at different stages of the environment-resource model with central waste treatment and subsequent recycling

A further decrease in the concentration ensues when the capital good is used in the consumption good process R_2. The capital good is physically used up in the course of time by deterioration. We have assumed that deterioration takes place continually with the production of the capital good: c_2 units are used up per unit of the capital good employed in R_2. Since k_2 units of the capital good are required for the production of one unit of the consumption good, $c_2 k_2$ units are used up. They contribute to gross emissions $S_2^b(s+3)$ for the production of one unit of the consumption good. Since these $c_2 k_2$ units of the capital good mix with other substances, the concentration of the resource falls again:

$$\kappa(S_2^b(s+3)) < \kappa(K(s+2)) < \kappa(s+1) \leqq 1. \qquad (7.7)$$

The decrease of the resource concentration is depicted in Fig. 7.3.

As shown in Fig. 7.1, the gross emissions $S_2^b(s+3)$ of process R_2 are mixed with those of process R_1 during centralized waste treatment. Since the exhaustible resource is not a factor of production in process R_1, it cannot be contained within the gross emissions $S_1^b(s+3)$ of this process. The resource concentration in the mixture of gross emissions of the two processes R_1 and R_2 is thus smaller than that of the gross emissions of process R_2 alone.

The concentration of the exhaustible resource in the gross emissions of the two production processes is the weighted average of the emissions of each of the processes. For this weighted average one obtains[4]

$$\kappa(S^B) = \frac{\kappa(S_2^b)\, S_2^b}{S_1^b + S_2^b} \tag{7.8}$$

since the emissions of the first process do not contain any of this substance. $\kappa(S_1^b)$ is thus 0.

Since the gross emissions of the two processes R_1 and R_2 also depend on the level at which they are run, we obtain from (7.8) by application of (7.7)

$$\begin{aligned} 0 = \kappa(S_1^b) &\leqq \kappa(S^b(s+3)) \leqq \kappa(S_2^b(s+3) \\ &< \kappa(K(s+2) < \kappa(s+1) \leqq 1 . \end{aligned} \tag{7.9}$$

We thus recognize from inequality (7.9) that the concentration of the resource decreases at each stage of production (cf. also Fig. 7.3).

It was assumed in Sect. 7.2.1 that only the gross emissions of the waste treatment process are used for recycling. For simplification we further make:

Assumption 7.4. The concentration of the resource in the emissions is not decreased any further by waste treatment, i.e.

$$\kappa(S_p^b) = \kappa(S_4^b). \tag{7.10}$$

This special assumption once again exemplifies a disadvantage of the centralization of waste treatment activities. It is a consequence of the mixing that the concentration of the resource is generally lowered. In practice it must thus be decided on a case to case basis whether the recovery of resources is effected centrally or separately for each emission.

What are the factors that determine whether recycling should be undertaken or not? A necessary condition for recycling is that

[4] To simplify the notation, time indices were omitted in formulae (7.8), (7.9), and (7.10).

the resource concentration $\kappa(S_4^b)$ within gross emissions must be so high that efficient production of consumption goods with recycling is possible. This implies (cf. the efficiency consideration in Chap. 6, Sect. 6.2) that the sum of the direct and indirect labor necessary for the production of one unit of the consumption good with technique T_2, with recycling applied, must be smaller or equal to the direct labor expended in producing one unit of the consumption good with technique T_1, i.e.[5]

$$l_1 \geqq l_2 + l_3 cK_; + l_6(\kappa(S_4^b))\, r_3\, ck_2 . \tag{7.11}$$

From inequality (7.11) we can thus determine a minimum concentration $\bar{\kappa}$ of the resources in the emissions for recycling to be technically efficient and therefore applied.[6] Let us now assume that the necessary condition (7.11) is met. It still does not follow with certainty that recycling is actually undertaken. This is because with recycling on the one hand, and extraction on the other, we have two means at our disposal with which to obtain resources for the production of the capital good. It follows from efficiency considerations that we shall choose that one of the two possible means of resource extraction which requires less effort. The following applies to our model if we adopt a short-run perspective that is confined to one period only. Recycling is not applied if

$$l_6(\kappa(S_4^b)) > l_5(\kappa(\bar{R})) .$$

Recycling is applied if

$$l_6(\kappa(S_4^b)) < l_5(\kappa(\bar{R})) .$$

Both procedures can be applied simultaneously if

$$l_6(\kappa(S_4^b)) = l_5(\kappa(\bar{R})) .$$

---

[5] Equation (7.11) can be directly derived from from Eq. (6.1). Equation (6.1) is a necessary condition that must be met in order for the production of consumption goods with scarce resources to be engaged in. Since according to assumption 7.3 the structure of the recycling process is identical to that of the extraction process, (7.11) is a direct implication of (6.1). These considerations, however, are only valid if $s_1 \leqq s_2$ (cf. Sect. 2.1.3, footnote 8).

[6] This minimum concentration is identical to the minimum concentration of the resource in the environment for the consumption good to be produced with the resource as an input.

Let us now remember that the following relationship holds (cf. Chap. 4)

$$l_5(\kappa(\bar{R})) \lesseqqgtr l_6(\kappa(S_4^b)) \quad \text{for} \quad \kappa(\bar{R}) \lesseqqgtr \kappa(S_4^b). \tag{7.12}$$

We have thus derived a condition for the application of recycling. Recycling is applied if the concentration of the resource in the emissions exceeds its concentration in the environmental sector.

In order to be able to decide from a long-run perspective that encompasses all T periods whether recycling is actually applied, we have to determine the temporal development of the resource concentration in the emissions and in the environment. In Fig. 7.4 the development of the resource concentration in the emissions and in the environment is depicted. We assumed: first, technique T_2 is only applied to a small extent because the resource must first be exploited and the quantities of the capital good necessary for the production of the consumption good must be manufactured. Then the production of the consumption good with technique T_2 is intensified until the factor input in resource extraction is so high that efficient production with technique T_2 is no longer possible (cf. Chap. 6). Resources are constantly extracted in this development of production; their concentration in the environmental

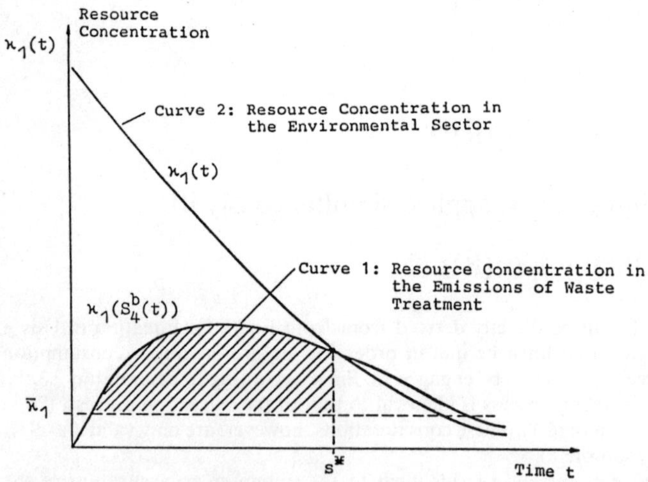

Fig. 7.4. Temporal development of resource concentration in the environmental sector and in the gross emissions of waste treatment

sector thus decreases. At the same time larger and larger quantities of resources are emitted, causing the concentration of the resources in the emissions to increase relative to the concentration in the environmental sector. There thus comes a time s* when the concentration of the resources in the emissions exceeds their concentration in the environmental sector (cf. Fig. 7.4). Efficient production will then involve recycling.

As a consequence of the temporal development of $\kappa(\overline{R})$ and $\kappa(S_4^b)$ it may, however, be useful to resort to recycling even before period s*. For if the application of recycling is postponed until period s*, then the gross emissions of waste treatment $S_4^b(t)$ $(t < s^*)$ will be returned to the environment without the resources contained therein being recovered. These resource quantities are thus diffusely distributed and their concentration may approach zero. However, since recycling requires a minimum concentration $\bar{\kappa}$ (cf. (7.9)), these resource quantities would be irrevocably excluded from recovery. The potential for ecological improvement that would thus be lost is represented by the striped area in Fig. 7.4.

These considerations show once more how important is the length of the planning horizon T for economic decisions. In Fig. 7.4, for instance, the following is true for every $T < s^*$:

$$\kappa(S_4^b(t)) < \kappa(t) \quad (t \leqq T \leqq s^*).$$

Recycling would thus be out of the question for these planning horizons.

Furthermore it might be conjectured that resource extraction from the environment is discontinued when its factor requirement is higher than in recycling, i.e. after period s*. This is not always the case, however, for if the resource is used exclusively for the production of the capital good – as in our model – the resource quantities extracted from the environment are linked to the capital stocks existing in the consumption and waste treatment sectors. Only through the deterioration of these capital goods will resources be set free in the course of time and made available for recycling. Now, if the rates of deterioration c_2 and c_4 are small, then for long periods only small amounts of resources will be available for recycling. Thus even if the concentration of the emissions allows an efficient use of recycling, the situation may occur that these quantities are not sufficient to cover the need for resources on the optimal path. This will be the case especially shortly after the introduction of technique T_2. Resources will thus be extracted from the environment even if the factor input in resource extraction is smaller than that in recycling.

We summarize our conclusions as follows. Whether recycling is applied or not cannot be decided by a comparison of the factor inputs in resource extraction and in recycling alone. For this criterion is rather coarse and geared towards short-run considerations. The exact timing of the introduction of recycling cannot be determined with this criterion alone. For a more precise determination of this timing, the temporal development of production must also be analyzed, as was done in Sect. 7.1 above.

7.2.3 Deposition as a Support for Recycling

Deposition in this context is the controlled collection of emissions from the production and waste treatment sectors in a spatially limited area. In Chap. 4, Sect. 4.2, we have dealt with the fact that it is possible to slow down the decrease of the resource concentration in the environmental sector, and thus to lower the increase in entropy. In the long-run this leads to a slower increase in the factor input into resource extraction. Up until now the possibility of a prolonged application of technique T_2 has not been considered. This was excluded by Assumption 5.2 (diffuse distribution of the emissions in the environment).

The form of the deposition of pollutants and waste materials in the environment can be of crucial importance in determining whether recycling is applied. For as we can see with the help of Fig. 7.4 recycling is only used after period s* for the winning of resources – if one applies the short-run criterion that we derived in the preceding Section. Now if the pollutants and waste materials that were emitted before period s* are deposited, the resources contained therein can also be recovered by recycling. In the case of a diffuse distribution of these emissions over the whole environmental sector, however, the resources contained therein are unavailable for recycling. Deposition can thus be seen as an option for the recovery of substances emitted in later periods.

The introduction of deposition into our total model is relatively simple. For this purpose we make:

Assumption 7.5. All emissions can be deposited, and are deposited in a spatially limited area u of the environmental sector. Emissions are thus collected in this restricted area in the course of time. The quantity of pollutants and waste materials deposited up until period s can therefore be calculated according to equation (cf. Fig. 7.2)

$$N^u(s) = \sum_{t=1}^{s-1} [S_p^n(t) + S_4^n(t)]. \tag{7.13}$$

The average concentration of the stock of exhaustible resources in the deposit can be estimated for each period by the quotient of the sum of the concentrations weighted with the respective quantities and the sum of these quantities, i.e. by formula (taking into consideration (7.10))

$$\kappa^u(s) = \frac{\sum_{t=1}^{s-1} \kappa[S_p^n(t))\,[S_p^n + S_4^n(t)]}{\sum_{t=1}^{s-1} [S_p^n(t) + S_4^n(t)]}. \tag{7.14}$$

With the average concentration of the resources in the deposit, $\kappa^u(s)$, we now have an analytical device at our disposal with which to derive a necessary condition for positive answers to the following questions:

1. Under what condition is it economically advantageous to set up a deposit with the objective later to recover resources from it?
2. From which time on is it advisable actually to recover the resources from this deposit?

In the preceding section it was shown that recycling can only be efficiently undertaken if the resource concentration is smaller in the environment than in the flow of emissions in at least one period during the planning horizon. Applying the same criterion to deposited emissions we can say that the recycling of resources from deposits can only be efficiently undertaken if the average concentration of resources in the deposit κ^u is higher than the concentration of the resource in the environment. In other words: If there is a period s, $s < T$, with

$$\kappa^u(s) > \kappa(s), \tag{7.15}$$

within the planning horizon, then it is advantageous to set up a deposit and to extract resources from it by recycling in period s. Like recycling, deposition can also be helpful in prolonging the period of application of a production technique which utilizes exhaustible resources. For if, due to low resource concentration, the exploitation of the natural deposits is no longer efficient, then there at least remains the stock of material collected as a precaution in the waste deposit, with which to apply the resource using technique a little longer. Once again, the real background for these considerations is as follows; deposition lowers the increase in entropy in the environmental sector.

In order properly to evaluate our results it has to be recalled, however, that we have formulated a necessary condition only. For the factor input into a deposit incurred in the transportation of waste materials, as well as in the upkeep of the deposit, has remained unconsidered. The transportation and the running of the deposit moreover create emissions that impair the environment and might possibly be cost-intensively treated.[7] Whether the deposition of waste materials is really economically advisable as a supportive measure of recycling thus depends not only on the resource concentration in the deposit relative to the resource concentration in the environment. It must also be evaluated from the perspectives of factor input (deposition costs) and environmental protection. In deposition, as in recycling, interrelationships between environmental protection and resource use must be taken into consideration.

[7] Effluents that have to be kept away from the groundwater may, for example, be emitted from the deposit (cf. SIMONS, HÄNSEL, and MESECK 1981 and JAEGER and WIENER 1981).

7.3 How Can We Avoid Wasting Low Entropy?[8]

In our view, a general aim of an ecologically oriented society should be not to waste low entropy. Examples of wasting low entropy are observed in the extraction of copper and in the use of natural gas.

a) During the past 200 years the concentration of copper in extraction sites has been reduced from 12% to 0.4%. Nevertheless the managements of respective mining firms admit that they consider the availability of copper and its concentration not as a restriction for their profit maximizing behavior, since it is economically feasible to extract copper down to 0.2% concentration, given the present prices (WAGENHALS 1984). This is a rather short-sighted reasoning, since prices can dramatically change in the course of time.
b) Helium is an exceptional substance, it has a unique nature.

> "Its most important properties are its inertness, low boiling point and lightness. However, it is this combination of properties in one substance that makes helium indispensable in many uses... Helium has been extensively used in the laboratory for lighter-than-air-craft, and for rockets." (EPPLER and LAVE 1980:617–8)

Helium is contained in natural gas. But in many instances it is not separated, but vented into the atmosphere, where its concentration is reduced from 3000 parts per million to 5 parts per million. Thus e.g. in the early seventies 1.4 billion cubic feet were vented in the USA per year (EPPLER and LAVE 1980: 618–9).

> "Since this failure to conserve helium means that we are increasing the cost of helium for future generations, some scientists suggest that the current government program (in the U.S.A., the authors) is immoral. Apparently, helium ought to be separated and stored from as many sources as would keep entropy from rising." (EPPLER and LAVE 1980:619).

Both examples show that the needs of future generations are not sufficiently taken care of. Using the model of intertemporal general equilibrium analysis (DEBREU 1959) one can partly explain this phenomenon by the circumstance that there are not

[8] This Section is based on Sect. 4 of FABER (1985).

sufficient future markets for these goods. But even if these did exist we could not expect that the situation would be much different from today. For assuming a social discount rate of 10%[9] and a period of 30 years for one generation, the present value of one Dollar profit in 2014 is equal to 0.06 Dollar and for two generations in 2044 equal to 0.003 Dollar. The reason for this result is the high discount rate of 10%, which in turn is due to the fact that only the living population is allowed to participate in the market process. If all future generations took part in the determination of prices the corresponding price system would, of course, be very different.

At first sight a remedy seems to be to recommend the government to decide over the realization of corresponding investments on the basis of a lower social rate of discount. However, every political party taking this proposal on its program would have a considerable disadvantage in winning the elections, since great amounts of taxes would have to be paid for purposes which would not benefit the present but future generations. Thus the opposition can gain votes from all those who want to pay less taxes. To solve this intricate problem we believe it is necessary first to convince the majority of the population that these kinds of future investments are an ethical obligation. Probably, only on the basis of such a general consensus will it be possible to carry through investments on behalf of future generations.[10]

---

[9] LIND (1982:9) reports that "... Resources for the Future sponsored a conference in 1966 on the appropriate rate of discount for evaluating water projects. The foremost economists and water policy analysists from the country were invited. After most of a day of debate without agreement on the conceptual basis for the choice of the discount rate, one of the conference's organizers posed the question. If you had to pick a single number, what would be your personal estimate of the appropriate rate? The answers ranged from a low of 2 to 4 percent, based on the real rate of return on long-tern government bonds, to 20 percent, based on the interest rate for short-term consumer credits."

[10] SPIEGEL (1971) "makes a contribution to the recent literature by taking note of the subjectivist position of Democritus (Greek philosopher of the fifth century B.C., the authors), among whose fragments survives a suggestion that improved levels of satisfaction can be achieved by reducing personal demand as well as by increasing supply." (LOWREY 1979:66)

7.4 Outlook

In the present Chapter we have analyzed the interdependencies between environmental protection and resource use within a single model. Two directions suggest itself for further investigations. First, we could generalize our approach by abandoning our special, linear structure of production in favor of a more general one. That having been done, a comparison of our results with those of the neoclassical theory of resource allocation (see e.g. DASGUPTA and HEAL 1979, DASGUPTA 1982 and SIEBERT 1983) would suggest itself.

A second would be to use our model for an investigation of concrete problems. Suitable fields of application would be the examples of atomic energy and of sewage sludges mentioned in the introductory remarks to Part II. In both problems resource extraction, recycling, deposition, and waste treatment have to be taken into consideration. In dealing with these areas we could draw on the experiences and results obtained in the context of our empirical studies of the water quality management of the BASF AG, Ludwigshafen FRG, of the State of Baden-Württemberg, FRG, and of Europe.

8 References

D'ARGE, R.C. (1972), "Economic Growth and the Natural Environment," in: A.V. KNEESE and B.T. BOWER (Eds.): Environmental Quality Analysis. Theory and Methods in the Social Sciences, Johns Hopkins University Press, Baltimore: 11–33.

ARROW, K. and M. KURZ (1970), Public Investment, the Rate of Return, and Optimal Fiscal Policy, Johns Hopkins University Press for Resources for the Future, Baltimore.

AYRES, R.U. and A.V. KNEESE (1969), "Production, Consumption, and Externalities," American Economic Review 59: 282–297.

BECKMANN, M.J. (1982), "Locational Aspects of Resource Depletion," in: W. EICHHORN and R. HENN (Eds.): Economic Theory of Natural Resources, Physica Verlag, Würzburg-Wien.

BENDER, D. (1976), Makroökonomik des Umweltschutzes (Macroeconomics of Environmental Protection), Vandenhoek und Ruprecht, Göttingen.

BENDER, E. and B. KONRAD (1977), "Sauerstoffkonzentrationsreglung in den Belebungsbecken der BASF-Kläranlage" (The Control of the Oxygen Concentration in the Activation Basins of the BASF Sewage Plant), Reglungstechnik 9:284–289.

BERNHOLZ, P. and F. BREYER (1984), Grundlagen der Politischen Ökonomie (Foundations of Politcal Economy), J.C.B. MOHR (Paul Siebeck), Tübingen.

BERNHOLZ, P. and M. FABER (1973), "Technical Productivity of Roundabout Processes and Positive Rate of Interest. A Capital Model with Depreciation and n-Period Horizon," Zeitschrift für die gesamte Staatswissenschaft 129:46–61.

BERNHOLZ, P., M. FABER, and W. REISS (1978), "A Neo-Austrian Two-Period Multisector Model of Capital," Journal of Economic Theory, Vol. 17, No. 1:38–50. (Reprinted in FABER 1986:98–112).

BLISS, C.J. (1975), Capital Theory and the Distribution of Income, North Holland Publishing Company, Amsterdam-Oxford.

BÖGE, W., M. FABER, and W. GÜTH (1982), "A Dynamic Game with Macroeconomic Investment Decisions under Alternative Market Structures," in: M. DEISTLER, E. FÜRST and G. SCHWÖDIAUER (Eds.): Games, Economic Dynamics, and Time Series Analysis, Physica-Verlag, Wien-Würzburg:227–250. (Reprinted in FABER 1986:229–258).

BREYER, F. and W. REISS (1981), "Probleme des Konzepts der Ressourcenausstattung" (Problems of the Concept of Resource Endowments), Zeitschrift für Wirtschafts- und Sozialwissenschaft 101: 83–98.

BUCHHOLZ, W. and D. CANSIER (1980), "Modelle ökologischen Wachstums" (Models of Ecological Growth), Zeitschrift für Wirtschafts- und Sozialwissenschaften 100:141–161.

BROWN, G.M and R.W. JOHNSON (1982), "Pollution Control by Effluent Charges: It Works in the Federal Republic of Germany. Why not in the U.S.," Natural Resource Journal 22:926–966.

BUCHWALD, K. and W. ENGELHARDT (Eds.) (1980), Handbuch für Planung, Gestaltung und Schutz der Umwelt. 3. Die Bewertung und Planung der Umwelt (Handbook for Planning, Design, and Protection of the Environment), BLV Verlagsgesellschaft, München-Bern-Wien.

BUPP, I.C. (1980), "Das nukleare Patt", in: R. STOBAUGH und D. YERGIN (Eds.): Energie-Report der Harvard Business School, Bertelsmann Verlag.

BULLINGER, M., G. RINCKE, A. OBERHAUSER, and R.B. SCHMIDT (1974), "Das Verursacherprinzip und seine Instrumente" (The Polluter-Pay-Principle and its Instruments), in: Beiträge zum Umweltschutz A 24 (Contributions to Environmental Protection A 24), Erich Schmidt Verlag, Berlin.

BUNDESREGIERUNG (1971), "Umweltprogramm der Bundesregierung 1971" (Environmental Program of the Federal Government 1971), Bundestagsdrucksache 6/2710, Heger, Bonn.

CHAPMAN, P.F. and F. ROBERTS (1983), Metal Resources and Energy, Monographs on Materials, Butterworth, London.

CLARK, C.W. (1976), Mathematical Bioeconomics, John Wiley & Sons, New York.

COMOLLI, P.M. (1977), "Pollution Control in a Simplified General Equilibrium Model with Production Externalities," Journal of Environmental Economics and Management 4:289–304.

DALY, H.E. (Ed.) (1973), Toward a Steady State Economy, W.H. FREEMAN and Company, San Francisco.

DASGUPTA, P.S. (1982), The Control of Resources, Basil Blackwell Publishers, Oxford.

DASGUPTA, P.S. and G.M. HEAL (1979), Economic Theory and Exhaustible Resources, J. NISBET & Co., Cambridge University Press, Cambridge.

DEBREU, G. (1959), Theory of Value, John Wiley & Sons, New York-London.

ENGELHARDT, H. und W.G. HALTRICH (1978), "Gewässerschutz in der BASF AG Ludwigshafen" (Water Protection in the BASF AG Ludwigshafen), Chemie-Ing.-Technik 50:510–518.

ENGELMANN, D. (1976), "Elektrische Ausrüstung der Kläranlage in Ludwigshafen" (Electrical Equipment of the Water Treatment Plant in Ludwigshafen), Chemie-Technik 5:163–168.

EPPLER, D. and L. LAVE (1980), "Helium: Investment in the Future," The Bell Journal of Economics 11:617–630.

FABER, M. (1979), Introduction to Modern Austrian Capital Theory, Lecture Notes in Economics and Mathematical Systems 167, Springer-Verlag, Berlin-Heidelberg-New York.

FABER, M. (1980), "Relationships Between Modern Austrian and Sraffa's Capital Theory," Zeitschrift für die gesamte Staatswissenschaft 136:617–629 (reprinted in FABER 1986:44–59).

FABER, M. (1981), "Modern Austrian Capital Theory and Orosel's Standard Neoclassical Analysis: A Reply," Zeitschrift für Nationalökonomie 41:157–176.

FABER, M. (1985), "A Biophysical Approach to the Economy. Entropy, Environment and Resources," in: W. VAN GOOL and J. BRUGGINK (Eds.): Energy and Time in Economic and Physical Sciences, Elsevier Science Publishers B.V., North-Holland.

FABER, M. (Ed.) (1986), Studies in Austrian Capital Theory, Investment and Time, Lecture Notes in Economics and Mathematical Systems 277, Springer-Verlag Berlin-Heidelberg-New York-London-Paris-Tokyo.

FABER, M., H. NIEMES, and G. STEPHAN (1983), Umweltschutz und Input-Output-Analyse. Mit zwei Fallstudien aus der Wassergütewirtschaft (Environmental Protection and Input Output Analysis. With Two Case-Studys in Water-Quality Management), J.C.B. MOHR (Paul Siebeck), Tübingen.

FABER, M. and J.L.R. PROOPS (1985), "Interdisciplinary Research Between Economists and Physical Scientists: Retrospect and Prospect," Kyklos, Vol. 38, Fasc. 4:599–616.

FABER, M. and J.L.R. PROOPS (1986), "Time Irreversibilities in Economics: Some Lessons from the Natural Sciences," in FABER 1986:294–316.

FABER, M. and G. STEPHAN (1981), "Neo-Austrian Characterization of Proportional Prices with Positive Rates of Interest Relative to the Growth Rate," in: FABER 1986:144–153.

FABER, M. and G. WAGENHALS (1987), "Towards a Long Run Balance Between Economics and Environmental Protection," in: SALOMONS, W. and FÖRSTNER, U. (Eds.): Environmental Impact and Management of Mine Tailings and Dredged Materials, Springer, Berlin-Heidelberg-New York-Tokyo (forthcoming).

FAHMI, H.P. (1981), "Woher stammen die Schwermetalle in Abwasser und Klärschlamm?" (Where Do the Heavy Metals in Effluents and Sludge Come From?), Chemische Rundschau 6, 34. Jg.:4.

FALK, G. and W. RUPPEL (1976), Energie und Entropie. Die Physik des Naturwissenschaftlers. Eine Einführung in die Thermodynamik (Energy and Entropy. The Physics of the Natural Scientist. An Introduction Into Thermodynamics.), Springer-Verlag, Berlin-Heidelberg-New York.

FISHER, A.C. and F.A. PETERSON (1976), "The Environment in Economics: A Survey," Journal of Economic Literature 14:1–33.

FISHER, A.C. (1981), Resource and Environmental Economics, Cambridge University Press, Cambridge-London-New York.

FORSTER, B.A. (1972), "A Note on Economic Growth and Environmental Quality," Swedish Journal of Economics 74:281–285.

FORSTER, B.A. (1975), "Optimal Pollution Control with a Nonconstant Exponential Rate of Decay," Journal of Environmental Economics and Management 2:1–6.

FORSTER, B.A. (1977), "Pollution Control in a Two-Sector Dynamic General Equilibrium Model," Journal of Environmental Economics and Management 4:305–312.

FØRSUND, F.R. (1975), "The Polluter Pay Principle and Transitional Period Measures in a Dynamic Setting," Swedish Journal of Economics 77:56–68.

GEORGESCU-ROEGEN, N. (1971), The Entropy Law and the Economic Process, Harvard University Press, Cambridge/Mass.

GEORGESCU-ROEGEN, N. (1975), "Energy and Economic Myths," Southern Economic Journal 41:347–379.

GEORGESCU-ROEGEN, N. (1978), "Energy Analysis and Economic Valuation," Southern Economic Journal 45:1023–1058.

GERTHSEN, Ch.G. (1964), Physik (Physics), Springer-Verlag Berlin-Heidelberg-New York.

GERYBADZE, A. (1982), Innovation, Wettbewerb und Evolution. Eine mikro- und makroökonomische Untersuchung (Innovation, Competition, and Evolution. A Micro- and Macroeconomic Investigation), J.C.B. Mohr (Paul Siebeck), Tübingen.

GLOBAL 2000 (1980), The Global 2000 Report to the President, U.S. Government Printing Office, Washington.

GRUVER, G.W. (1976), "Optimal Investment in Pollution Control Capital in a Neoclassical Growth Context," Journal of Environmental Economics and Management 3:165–177.

HADLEY, G. (1964), Nonlinear and Dynamic Programming Reading, Addison-Wesley Publishing Company, Inc., Reading/Mass.

HALTRICH, W. (1978), "Erfolge des Gewässerschutzes bei der BASF" (Water Protection Successes in the BASF), in: Forum Städte-Hygiene 29:265–266.

HEAL, G. (1975), "Economic Aspects of Nature Resource Depletion," in: D.W. PEARCE and J. ROSE (Eds.): The Economics of Natural Resources, John Wiley & Sons, New York: 118–139.

HERFINDAHL, O.C. (1967), "Depletion and Economic Theory," in: GAFFNEY (Ed.): Resources and Taxation, University of Wisconsin Press, Madison-Milwaukee-London.

HIRSHLEIFER, J. (1972), Investment, Interest and Capital, Englewood Cliffs, New York.

HOEL, M. (1980), Extraction of an Exhaustible Resource under Uncertainty, Oelschlager, Gunn, and Hain, Königstein/Ts.

HOTELLING, H. (1931), "The Economics of Exhaustible Resources," Journal of Political Economy, 39:137–175.

JAEGER, K. (1980), "Ansätze zu einer ökonomischen Theorie des Recycling" (Approaches to an Economic Theory of Recycling), in: H. SIEBERT (Ed.): Erschöpfbare Ressourcen (Exhaustible Resources), Schriften des Vereins für Socialpolitik, N.F. 108, Duncker & Humblot, Berlin: 149–174.

JÄGER, B. und K. WIENER (1981), "Möglichkeiten der Sickerwasser-Minimierung in abgeschlossenen Deponien" (Possibilities of Minimizing Water Seeping Through the Ground in Closed Dumps), in Dokumentation – 5. Europäisches Abwasser- und Abfallsymposium, EAS – 22.-26. Juni 1981, München; Hrsg. EAS-Sekretariat, c/o Abwassertechnische Vereinigung e.V. (ATV), St. Augustin: 689–699.

JAKSCH, H.J. (1975 a), "Die Mehrergiebigkeit längerer Produktionsumwege in einem linearen Vielsektorenmodell" (The Superiority of Roundabout Methods in a Linear Multisector Model), Zeitschrift für die gesamte Staatswissenschaft 131: 92–105.

JAKSCH, H.J. (1975 b), "Effizienz und Minderschätzung künftiger Bedürfnisse in einem linearen Vielsektorenmodell" (Efficiency and Impatience to Consume in a Linear Multisector Model), in: O. BECKER and R. RICHTER, (Eds.): Dynamische Wirtschaftsanalyse (Dynamic Economic Analysis), J.C.B. MOHR (Paul Siebeck), Tübingen: 87–104.

KEELER, E. and M. SPENCER, and R. ZECKHAUSER (1971), "The Optimal Control of Pollution," Journal of Economic Theory 4: 19–34.

KEMP, M.C. und N.V. LONG (1980), "The Optimal Order of Exploitation of Deposits of an Exhaustible Resource," in: H. SIEBERT (Ed.): Erschöpfbare Ressourcen (Exhaustible Resources), Schriften des Vereins für Socialpolitik, N.F. 108, Duncker & Humblot, Berlin: 301–334.

KOOPMANS, T.C. (1951), "Analysis of Production as an Efficient Combination of Activity," in: T.C. KOOPMANS (Ed.): Activity Analysis of Production and Allocation, J. Wiley & Sons, New York-London.

KOOPMANS, T.C. (1957), "Allocation of Resources and the Price System," in Three Essays on the State of Economic Science, McGraw-Hill, New York-Toronto-London: 1–126.

KOOPMANS, T.C. (1960), "Stationary Ordinal Utility and Impatience," Econometrica 28: 287–309.

KOOPMANS, T.C. (1979), "Economics Among the Sciences," American Economic Review 69: 1–13.

KREEB, K.H. (1979), Ökologie und menschliche Umwelt (Ecology and the Human Environment), G. Fischer, Stuttgart-New York.

KÜMMEL, R. (1980), Growth Dynamics of the Energy Dependent Economy, Mathematical Systems in Economics, Vol. 54, Athenäum, Hain, Scriptor, Hanstein, Königstein/Ts.

LEONTIEF, W. (1953), Studies in the Structure of the American Economy, Oxford University Press, New York.

LIND, R. and others (1982), Discounting for Time and Risk in Energy Policy, the John Hopkins University Press for Resources for the Future, Baltimore, London.

LOTKA, A. (1956), Elements of Mathematical Biology, Dover.

LOWREY, S.T. (1979) "Recent Literature on Ancient Greek Economic Thought," Journal of Economic Literature 17:65–86.

MÄLER, K.-G. (1974), Environmental Economics: A Theoretical Inquiry, John Hopkins University Press, Baltimore-London.

MAGAT, W.A. (1978), "Pollution Control and Technological Advance: A Dynamic Model of the Firm," Journal of Environmental Economics and Management 5:1–25.

MAIER, G. (1984), Rohstoffe und Innovationen. Eine dynamische Untersuchung (Resources and Innovation. A Dynamic Investigation), Mathematical Systems in Economics 68, Athenäum, Hain, Scriptor, Hanstein, Königstein/Ts.

MALINVAUD, E. (1953), "Capital Accumulation and Efficient Allocation of Resources," Econometrica 21:233–268.

MALLE, K.G. (1977), "Flankierende Maßnahmen zur Reinigung der Abwässer der BASF" (Actions in Support of the Purification of the Effluents of the BASF), Zeitschrift für Wasser- und Abwasserforschung 5:151–155.

MALLE, K.G. (1978), "Die Halbierung der Restfracht" (The Halving of the Rest-Freight), Wasser, Luft und Betrieb 22.

MIROWSKI, P. (1984), "Physics and the Marginalist Revolution," Cambridge Journal of Economics 8:361–379.

MÖLLER, H., R. OSTERKAMP, and W. SCHNEIDER (1981), Umweltökonomik. Ein Überblick zur Einführung in die ökonomische Analyse von Umweltproblemen (Environmental Economics. A Survey Introducing into the Economic Analysis of Environmental Problems.), Athenäum-Verlag, Königstein/Ts.

MÖLLER, H., R. OSTERKAMP, and W. SCHNEIDER (Eds.) (1982), Umweltökonomik. Beitrag zur Theorie und Politik (Environmental Economics. A Contribution to Theory and Policy), Athenäum, Hain, Scriptor, Hanstein, Königstein.

MOLL, W.L.H. (1978), Taschenbuch für Umweltschutz I: Chemische und technologische Informationen (Pocket Book for Environmental Protection I: Chemical and Technology Information), Steinkopff Verlag, UTB 511, 2. Auflage, Darmstadt.

MUROTA, T.I. (1984), "Heat Economy of the Water Planet Earth: An Entropy Analysis and the Water-Soil Matrix Theory," Hitstsubashi Journal of Economics 25:161–172.

NIEMES, H. (1978), "Versuch einer ökonomischen Bewertung von Abwasserkläranlagen am Beispiel der Großkläranlage der BASF AG in Ludwigshafen/Rhein" (An Attempt at an Economic Evaluation of Effluent Purification Systems), Chemiker Zeitung 5:179–199.

NIEMES, H. (1981), Umwelt als Schadstoffempfänger – Die Wassergütewirtschaft als Beispiel (The Environment as a Recipient of Pollutants – The Example of Water Quality Management), Schriften zur Umwelt und Ressourcenökonomie, J.C.B. MOHR (Paul Siebeck), Tübingen.

OCKER, A. (1974), "Wachstum und Umwelt: Ein Modell optimaler Kontrolle" (Growth and Environment: A Model of Optimal Control), Weltwirtschaftliches Archiv 110:609–622.

PLOURDE, C.G. (1970), "A Simple Model of Replenishable Natural Resource Exploitation," American Economic Review 60:518–522.

PLOURDE, C.G. (1972), "A Model of Waste Accumulation and Disposal," Canadian Journal of Economics 5:120–125.

POHLE, R. (1981), "Wiederverwertung in Nürnberg" (Recycling in Nürnberg), Kommunalwirtschaft 1:16–19.

PRIGOGINE, I. and I. STENGERS, (1980), Dialog mit der Natur. Neue Wege naturwissenschaftlichen Denkens (Dialogue With Nature. New Ways of Natural Scientific Thinking), Piper & Co., München-Zürich.

PROOPS, J. L. R. (1985), "Thermodynamics and Economics: from Analogy to Physical functioning", in W. v. GOOL and J. BRUGGINK (Eds.), North-Holland, Amsterdam.

PROOPS, J. L. R., (1987), "Entropy, Information and Confusion in the Social Sciencies", The Journal of Interdisciplinary Economics, 1:225–242.

RAMSAY, F. (1928), "A Mathematical Theory of Saving," Economic Journal 38:543–559.

Rat von Sachverständigen für Umweltfragen (1976), "Umweltprobleme des Rheins" (Environmental Problems of the Rhine), 3. Sondergutachten, W. KOHLHAMMER, Stuttgart-Mainz.

Rat von Sachverständigen für Umweltfragen (1978), Umweltgutachten 1978 (Report on the Environment), Bundestagsdrucksache 98/1938, Heger, Bonn.

REISS, W. (1979), "Substitution in a Neo-Austrian Model of Capital", Zeitschrift für Nationalökonomie 39:33–52.

REISS, W. (1981), Umwegproduktion und Positivität des Zinses. Eine neo-österreichische Analyse (Roundaboutness and the Positivy of the Rate of Interest. A Neo-Austrian Analysis), Duncker & Humblot, Berlin.

REISS, W. and M. FABER (1982) "Own Rates of Interest in a General Model of Capital," in: M. DEISTLER, E. FÜRST, and G. SCHWÖDIAUER (Eds.): Games, Dynamics, and Time Series Analysis, Physica-Verlag, Wien-Würzburg:194–209. (Reprinted in FABER 1986:113–131).

SAMUELSON, P.A., (1983), "Rigorous Observational Positivism; Klein's Envelop. Aggregation; Thermodynamics and Economic Isomorphismus", in: F.G. ADAMS and B.G. HICKMAN (Eds.): Global Econometrics. Essays in Honor of Lawrence R. KLEIN, MIT Press, Cambridge/Mass.-London.

SCHENKEL, W. (1980), "Abfallbeseitigung und Rohstoffrückgewinnung" (Waste Material Deposition and the Recuperation of Resources), in: H.A. OOMATIA (Ed.): 3. Internationaler Kongreß "Energie und Umwelt. Technologische und wirtschaftliche Perspektiven" vom 11. bis 13.02.1980 in Düsseldorf, Vulkan Verlag, Essen:171–175.

SEN, A.K. (1970), Collective Choice and Social Welfare, Oliver & Boyd, Edinburgh.

SHANNON, C.E. (1948), "A Mathematical Theory of Communication," Bell System Technical Journal 27:379–423, 623–656. (Reprinted in C.E. SHANNON und W. WEAVER, Mathematical Theory of Communication, The University of Illinois Press, Urbana, Illinois 1949.)

SHELL, K. (1967), "Optimal Programs of Capital Accumulation for an Economy in which there is Exogenous Technical Change," in: K. SHELL (Ed.): Essays on the Theory of Optimal Economic Growth, MIT-Press, Cambridge/Mass.: 1–30.

SIEBERT, H. (1978), Ökonomische Theorie der Umwelt, J.C.B. MOHR (Paul Siebeck), Tübingen. (English Translation (1981), Economics of Environment, D.C. HEATH, Lexington).

SIEBERT, H. (1983), Ökonomische Theorie natürlicher Ressourcen (Economic Theory of Natural Resources), J.C.B. MOHR (Paul Siebeck), Tübingen.

SIMONS, H., W. HÄNSEL, and H. MESECK (1981), "Beständigkeit von Deponiedichtungen aus Ton gegen Sickergewässer" (Continualness in Dump Seals Made out of Clay), in Dokumentation – 5. Europäisches Abwasser- und Abfallsymposium, EAS-Sekretariat, c/o Abwassertechnische Vereinigung e.V. (ATV), St. Augustin: 673–688.

SMITH, V.L. (1972), "Dynamics of Waste Accumulation: Disposal Versus Recycling," Quarterly Journal of Economics 86:601–616.

SMITH, V.L. (1977), "Control Theory Applied to Natural and Environmental Resources," Journal of Environmental Economics and Management 4:1–24.

SOLOW, R. (1974), "The Economics of Resources or the Resources of Economics," American Economic Review 64:1–14.

SPIEGEL, H.W. (1971), The Growth of Economic Thought, Prentice Hall, Englewood Cliffs-London.

STEPHAN, G. (1980), Preise und Zinsen in Modellen mit unendlichem Horizont. Beziehungen zwischen der neoklassischen Aktivitätsanalyse und der neoösterreichischen Kapitaltheorie (Prices and Interest in Infinite Horizon Models. Relationships Between Neoclassical Activity Analysis and Neo-Austrian Capital Theory), R.G. FISHER, Frankfurt a. M.

STEPHAN, G. (1983), "Roundaboutness, Nontightness, and Malinvaud Prices in Multisector Models with Infinite Horizon," Zeitschrift für die gesamte Staatswissenschaft 131:660–677 (reprinted in FABER (1986):154–172).

STEPHAN, G. (1986), "A Neo-Austrian Approach to the Open-Endedness of the Future: An Overview," in FABER (1986):132–143.

STEPHAN, G. (1988), Emmission Standards, Substitution of Techniques and Long-run Adjustment: A Computable Equilibrium Approach, Forthcoming in: Microeconomic Studies (W. GÜTH and H.W. SINN Eds.) Springer-Verlag Berlin-Heidelberg-New York-London-Paris-Tokyo.

STEPHENS, J.K. (1976), "A Relatively Optimistic Analysis of Growth and Pollution in a Neoclassical Framework," Journal of Environmental Economics and Management 3:85–96.

STUMPF, H. and A. RIECKERS (1976), Thermodynamik (Thermodynamics), Vols. 1 und 2, Vieweg, Braunschweig.

THOSS, R., P. BRASSE, W. BURGBACHER, H.-P. DÖLLEKES, K.J. WIIK, and N. WIPPO (1975), "Umweltbilanzen und ökologische Lastpläne für Regionen" (Environmental Ballances and Ecological Load Plans for Regions), Teilprojekt K1, in Wissenschaftlicher Arbeitsbericht 1974 bis 1975 des SFB 26, Universität Münster.

VICTOR, P.A. (1972) Pollution: Economy and Environment, George Allen & Unwin Ltd., London.

VOGT, W. (1981), Zur intertemporalen wohlfahrtsoptimalen Nutzung knapper natürlicher Ressourcen (On the Intertemporal Optimal Use of Scarce Natural Resources), Schriftenreihe Umwelt und Ressourcenökonomie, J.C.B. MOHR (Paul Siebeck), Tübingen.

WAGENHALS, G. (1984), The World Copper Market: Structure and Econometric Model, Lecture Notes in Economics and Mathematical Systems 233, Springer-Verlag, Berlin-Heidelberg-New York.

WITTMANN, W. (1986), Produktionstheorie (Production Theory), Springer-Verlag, Berlin-Heidelberg-New York.

WODOPIA, F.J. (1986), Intertemporale Produktionsentscheidungen: Theorie und Anwendung auf die Elektrizitätswirtschaft (Intertemporal Production Decisions: Theory and Application on the Electricity Sector), Mathematical Systems in Economics 102, Athenäum, Hain, Hanstein, Königstein/Ts.

ZWINTSCHER, K. (1973), "Social Costs in der öffentlichen Trinkwasserversorgung" (Social Costs in the Public Supply of Drinking Water), Veröffentlichung des Institutes für Wasserforschung Nr. 17, Dortmund.

9 Subject Index